CCEA A2
CHEMISTRY EXAM PRACTICE

COLOURPOINT EDUCATIONAL

© Alyn McFarland, Nora Henry and Colourpoint Creative Ltd 2021

Print ISBN: 978 1 78073 254 1
eBook ISBN: 978 1 78073 308 1

First Edition
Second Impression 2022

Layout and design: April Sky Design
Printed by: GPS Colour Graphics, Belfast

All rights reserved. No part of this publication may be reproduced, stored in a retrieval system or transmitted in any form or by any means, electronic, mechanical, photocopying, scanning, recording or otherwise, without the prior written permission of the copyright owners and publisher of this book.

Copyright has been acknowledged to the best of our ability. If there are any inadvertent errors or omissions, we shall be happy to correct them in any future editions.

Colourpoint Educational
An imprint of Colourpoint Creative Ltd
Colourpoint House
Jubilee Business Park
21 Jubilee Road
Newtownards
County Down
Northern Ireland
BT23 4YH

Tel: 028 9182 0505
E-mail: sales@colourpoint.co.uk
Web site: www.colourpoint.co.uk

The Authors

Dr Alyn G McFarland has been teaching GCSE and GCE A-level Chemistry for 26 years and was Head of Chemistry in a large grammar school for 14 years. He has been writing textbooks, revision guides and workbooks for GCSE Chemistry and GCE A-level Chemistry for different examination boards for over 10 years. Dr McFarland is a senior examiner at both levels for an examination board and also contributes to the PGCE course for Science/Chemistry students on a part-time basis.

Nora Henry is a teacher at a Belfast grammar school and a part-time tutor for a university education department. She works for an examining body as Principal Examiner for GCSE Chemistry, Reviser for A level Chemistry and Reviser for A Level Life and Health Sciences. In addition to this text, she has written around 30 textbooks, workbooks and study guides for GCSE and A Level.

Publisher's Note: This book has been written to help students preparing for the A2 Level Chemistry specification from CCEA. While Colourpoint Educational and the author have taken every care in its production, we are not able to guarantee that the book is completely error-free. Additionally, while the book has been written to closely match the CCEA specification, it is the responsibility of each candidate to satisfy themselves that they have fully met the requirements of the CCEA specification prior to sitting an exam set by that body. For this reason, and because specifications change with time, we strongly advise every candidate to avail of a qualified teacher and to check the contents of the most recent specification for themselves prior to the exam. Colourpoint Creative Ltd therefore cannot be held responsible for any errors or omissions in this book or any consequences thereof.

Health and Safety: This book describes practical tasks or experiments that are either useful or required for the course. These must only be carried out in a school setting under the supervision of a qualified teacher. It is the responsibility of the school to ensure that students are provided with a safe environment in which to carry out the work. Where it is appropriate, they should consider reference to CLEAPPS.

CONTENTS

Introduction 4

A2 1 Further Physical and Inorganic Chemistry

4.1   Lattice Enthalpy 7

4.2   Enthalpy, Entropy and Free Energy 17

4.3   Rates of Reaction 26

4.4   Equilibrium.. 36

4.5   Acid-Base Equilibria 46

4.6   Isomerism 67

4.7   Aldehydes and Ketones 73

4.8   Carboxylic Acids 86

4.9   Derivatives of Carboxylic Acids 94

4.10   Aromatic Chemistry 108

A2 2 Analytical, Transition Metals, Electrochemistry and Organic Nitrogen Chemistry

5.1   Mass Spectrometry.. 119

5.2   Nuclear Magnetic Resonance Spectroscopy................................ 128

5.3   Volumetric Analysis 136

5.4   Chromatography 151

5.5   Transition Metals.. 159

5.6   Electrode Potentials... 177

5.7   Amines... 189

5.8   Amides.. 203

5.9   Amino Acids 209

5.10   Polymers 217

5.11   Chemistry in Medicine 225

Introduction

This workbook for CCEA A2 Chemistry is divided into the two content units A2 1 and A2 2. The topics in each A2 unit are comprised of a series of questions including multiple choice questions and structured questions and where appropriate practical questions and calculations. Topics are subdivided to assist your revision. All answers are provided online with worked solutions to calculations.

A2 examinations

The A2 examinations for CCEA Chemistry comprise A2 1 (Further Physical and Organic Chemistry), A2 2 (Analytical, Transition Metals, Electrochemistry and Organic Nitrogen Chemistry) and A2 3 (Further Practical Chemistry).

A2 1 and A2 2 are both worth 110 raw marks and consist of 10 multiple choice questions worth 1 mark each and 100 marks of structured questions.

A2 3 consists of two components. A2 3 Booklet A is a practical examination carried out in your school laboratory and is worth 30 raw marks. A2 3 Booklet B is a practical theory examination and is a timetabled examination worth 60 raw marks. A2 3 can cover any practical aspects of all the content in A2 1 and A2 2 and questions are found throughout the topics in the workbook.

Multiple choice questions will always include distractors, so read all answers carefully. Don't spend too much time on the multiple choice questions as they are only worth 1 mark each so better to come back to the ones you need to think about at the end if you have time. There are no multiple choice questions in A2 3.

A2 1 and A2 2 contain "quality of written communication" questions which will assess your ability to write coherently in proper sentences with correct spelling, punctuation and grammar. There are two of these questions on A2 1 and A2 2 and they are clearly labelled. Note that all A2 examination papers will contain questions on synoptic content from AS Chemistry.

Using past papers

Be aware that raw marks in the individual units are converted to uniform marks (UMS) and the grade boundaries change from module to module but these are published on the CCEA microsite so always check what grade you would have achieved in the unit you tried if you are using past papers. Legacy units (from previous specifications) are also useful for revision but be careful as some topics have moved to a different unit or may have removed. Check with the latest specification or ask your teacher if you are unsure.

Command words

Command words are important so make sure you read the questions carefully. "State and explain" means you should state a trend or pattern and then explain why this occurs. "Suggest" is often used if the question is asking you to apply your knowledge from the specification in an unfamiliar context. "Calculate" will be used where you have to carry out a calculation and show the steps in your calculation. Calculation are marked based on errors made with each error losing a mark. Errors are also carried forward so make sure you show all steps clearly as some marks may still be obtained even if you make a mistake. "Name" means you would provide a name and not a formula. Be careful with organic nomenclature as errors in using commas and dashes are penalised.

CCEA support documents

CCEA provide guidance on "Clarification of terms", "Acceptable colour changes and observations" as well as AS and A2 "Practical support documents". These should be adhered to carefully. Errors in definitions are penalised by each error.

Colours which are separated by a solidus (/) mean alternatives. Only one of the alternatives should be given. For example the colour of a solution of bromine is yellow/orange/brown so only one of these colours should be used in an observation question. Colours which are separated by a dash must always include the dash such the flame test colour for copper(II) ions which is blue-green or green-blue. Both colours should be provided and should be separated by a dash (never a solidus). Make sure you apply this to colour changes too. The practical document gives suggested methods for all practical activities detailed in the specification.

Level of precision in calculations

Many calculations will include an instruction to give your answer to a specified number of significant figures or decimal places. This is only for the final answer given and it is often good practice to work through the calculation to a number of decimal places or significant figures one higher than the level requested for the final answer and round appropriately at the end. In some calculations you will be asked to give your answer to an appropriate level of significant figures. You should check the numbers of significant figures for each piece of data provided in the question and give your answer to same level of precision as the least precise piece of data. For example, a question with all the data provided to 3 significant figures would require an answer to 3 significant figures. However, a calculation with most figures to 3 significant figures but having one to 2 significant figures would require an answer to 2 significant figures. RAMs/RFMs/RMMs and balancing numbers in equations do not affect the number of significant figures in your answer so these can be ignored.

Drawing diagrams of apparatus

Diagrams, when asked for, should be cross-sectional and show a free flow of the liquids and gases in the apparatus with no blockages caused by line across the flow. Draw a two-dimensional representation of the apparatus and ensure you label all appropriate apparatus. Include heat where required and include labels for "water in" and "water out" in reflux and distillation.

Finally

Questions will address all the assessment objectives within the specification. Read the stem carefully as often there is information which will assist you in answering the questions which follow. Work through the paper and be aware of time. Check you have not missed any pages as it does happen more often than you think.

The mark scheme (the answers) for this workbook is available online. Visit www.colourpointeducational.com and search for *Chemistry Exam Practice for CCEA A2 Level*. The page for this book will contain instructions for downloading the mark scheme. If you have any difficulties please contact Colourpoint.

Good luck!

Unit A2 1:
Further Physical and Inorganic Chemistry

4.1 Lattice Enthalpy

1 Which one of the following represents the enthalpy of formation of magnesium oxide?

A   Mg(s) + O(g) → MgO(s)

B Mg^{2+}(g) + O^{2-}(g) → MgO(s)

C   Mg(s) + ½O_2(g) → MgO(s)

D   Mg(g) + ½O_2(g) → MgO(s)   [1]

2 The lattice enthalpy of calcium chloride is +2237 kJ mol^{-1}. The enthalpy of hydration of calcium ions is −1650 kJ mol^{-1} and of chloride ions is −364 kJ mol^{-1}. What is the enthalpy of solution of calcium chloride?

A   −223 kJ mol^{-1}

B   −141 kJ mol^{-1}

C   +223 kJ mol^{-1}

D   +951 kJ mol^{-1} [1]

3 Give the definitions of the following:

(a) First ionisation energy _____

_____ [2]

(b) Standard enthalpy of formation _____

_____ [2]

(c) Standard enthalpy of atomisation _____

_____ [2]

(d) Lattice enthalpy _____

_____ [2]

A2 1: FURTHER PHYSICAL AND INORGANIC CHEMISTRY

4 The information below relates to the formation of caesium chloride, CsCl. Caesium is a solid at room temperature and pressure.

	ΔH⦵ / kJ mol⁻¹
First ionisation energy of caesium	+380
Enthalpy of atomisation of caesium	+78
Enthalpy of formation of caesium chloride	−433
Enthalpy of atomisation of chlorine	+122
First electron affinity of chlorine atoms	−364

(a) Write equations, including state symbols, for the reactions which would have enthalpy changes equal to the following:

(i) The first ionisation energy of caesium

_____ [1]

(ii) The enthalpy of formation of caesium chloride

_____ [1]

(iii) The first electron affinity of chlorine

_____ [1]

(iv) The lattice energy of caesium chloride

_____ [1]

(b) (i) Complete the missing levels of the Born-Haber cycle for caesium chloride below. Include state symbols.

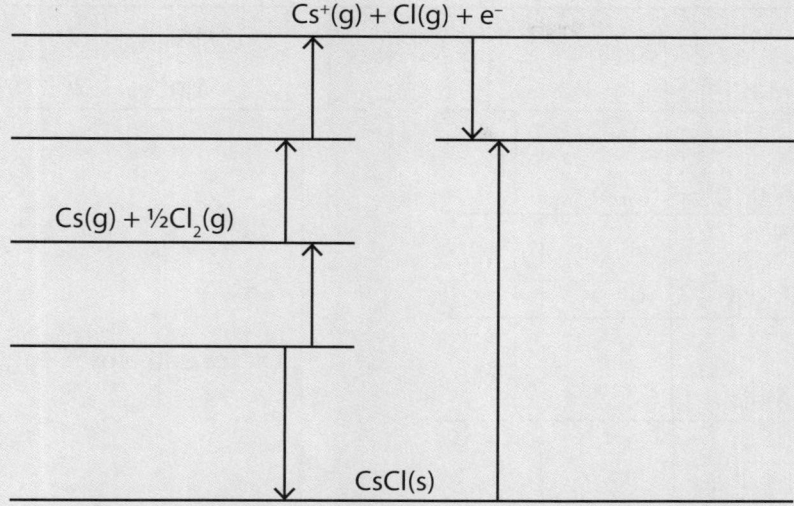

[3]

(ii) Using the constructed Born-Haber cycle, or any other method, calculate the lattice enthalpy of caesium chloride.

_____

_____

_____

_____ [2]

(c) The lattice enthalpies for sodium chloride, potassium chloride and rubidium chloride are +776, +710 and +685 kJ mol^{-1} respectively. Suggest why there is a difference in these results compared with the value you calculated for caesium chloride in (b)(ii).

_____

_____

_____

_____ [2]

5 The Born-Haber cycle shown below is for magnesium chloride.

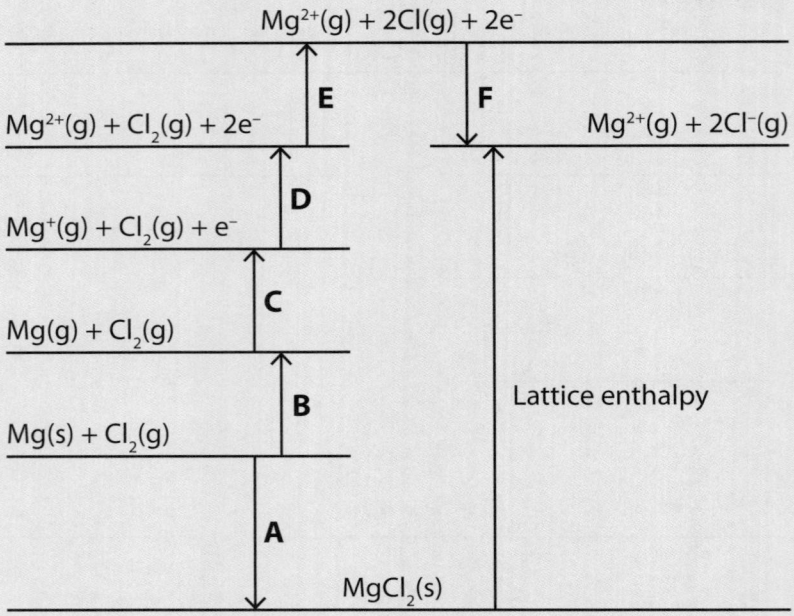

(a) Write the letters beside the terms below which are represented on the diagram above.

Standard enthalpy of formation of magnesium chloride (−642 kJ mol⁻¹) _____

Second ionisation energy of magnesium (+1450 kJ mol⁻¹) _____

Standard enthalpy of atomisation of magnesium (+150 kJ mol⁻¹) _____

Standard bond dissociation enthalpy of chlorine (+242 kJ mol⁻¹) _____

First ionisation energy of magnesium (+736 kJ mol⁻¹) _____

First electron affinity of chlorine (−364 kJ mol⁻¹) _____ [2]

(b) Calculate the lattice enthalpy of magnesium chloride using the data given above.

_____

_____

_____

_____

_____ [3]

(c) The enthalpy of solution of magnesium chloride is −165 kJ mol⁻¹. The hydration enthalpy of the magnesium ion is −1891 kJ mol⁻¹. Using this information and the value calculated in (b), calculate the hydration enthalpy of the chloride ion.

_____

_____

_____

_____ [3]

(d) Explain, using a diagram, why energy is released when magnesium and chloride ions become hydrated.

_____

_____

_____

_____ [4]

6 The diagram below shows a Born Haber cycle for silver(I) fluoride.

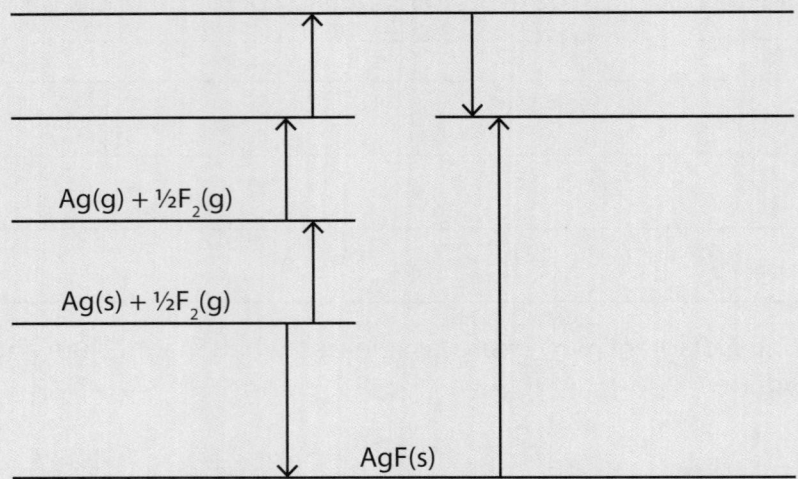

(a) Complete the cycle by filling in the missing spaces with the particles involved. [3]

(b) The information in the table shows the enthalpy changes associated with this Born Haber cycle.

Enthalpy change	$\Delta H^\ominus$ / kJ mol^{-1}
Enthalpy of formation of silver(I) fluoride	−203
Enthalpy of atomisation of silver	+286
First ionisation energy of silver	+730
Enthalpy of atomisation of fluorine	+79
Lattice enthalpy of silver(I) fluoride	+943

Calculate the first electron affinity for fluorine using the information in the table.

_____

_____

_____

_____ [2]

(c) The diagram below shows the enthalpy changes which occur when silver(I) fluoride dissolves in water.

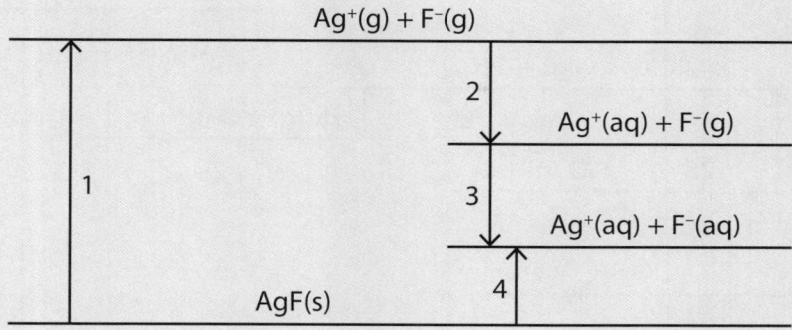

(i) Name the enthalpy changes labelled 1, 2, 3 and 4.

1 _____

2 _____

3 _____

4 _____ [4]

(ii) The table below shows the values for the enthalpy changes 1, 2 and 3. Calculate a value for enthalpy change 4.

Enthalpy change	Value / kJ mol^{-1}
1	+943
2	−464
3	−457

_____

_____

_____ [2]

7 The Born-Haber cycle for potassium oxide is shown below.

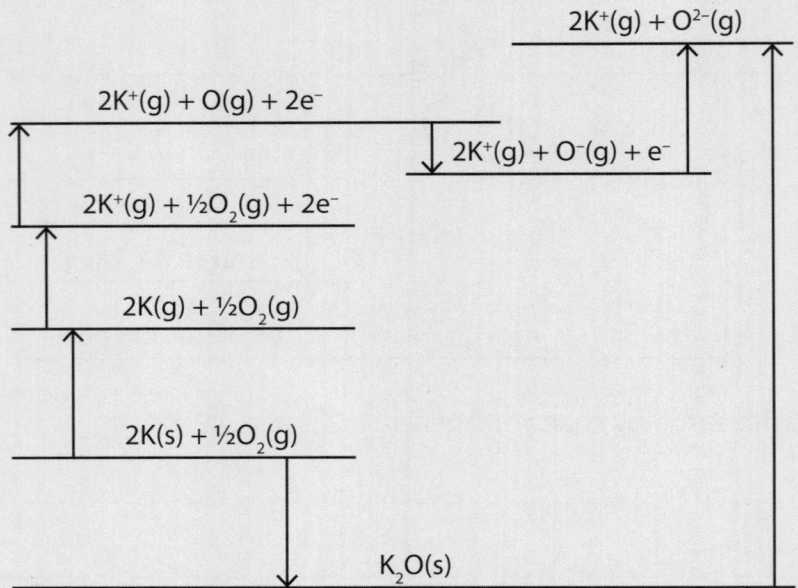

The table below shows the enthalpy changes associated with this Born-Haber cycle.

Enthalpy change	ΔH⦵ / kJ mol⁻¹
Enthalpy of formation of potassium oxide	−561
Enthalpy of atomisation of potassium	+90
Bond dissociation enthalpy of oxygen	+496
First electron affinity of oxygen	−142
Second electron affinity of oxygen	+844
Lattice enthalpy of potassium oxide	+2531

(a) Label each of the arrows with the value of the enthalpy changes for the reaction shown in the cycle. [5]

(b) Calculate the first ionisation energy of potassium.

_____

_____

_____

_____ [3]

4.1 LATTICE ENTHALPY

(c) Some of the values of the enthalpy changes are different in the Born-Haber cycle for sodium oxide. The table below shows the values.

Enthalpy change	ΔH⦵ / kJ mol⁻¹
Enthalpy of formation of sodium oxide	−416
Enthalpy of atomisation of sodium	+108
Bond dissociation enthalpy of oxygen	+496
First electron affinity of oxygen	−142
Second electron affinity of oxygen	+844
Lattice enthalpy of sodium oxide	to be calculated
First ionisation energy of sodium	+500

(i) Explain why the enthalpy of atomisation of sodium is greater than the enthalpy of atomisation of potassium.

_____
_____
_____ [2]

(ii) Explain why the first ionisation energy of sodium is greater than the first ionisation energy of potassium (the value calculated in (b)).

_____
_____
_____ [2]

(iii) Calculate the lattice enthalpy of sodium oxide.

_____
_____
_____
_____ [3]

A2 1: FURTHER PHYSICAL AND INORGANIC CHEMISTRY

8 Part of the Born-Haber cycle for magnesium oxide is shown below. The values for the enthalpy changes are given in the table.

Enthalpy change	ΔH$^\ominus$ / kJ mol^{-1}
Enthalpy of atomisation of magnesium	+150
First ionisation energy of magnesium	+736
Second ionisation energy of magnesium	+1450
Enthalpy of atomisation of oxygen	+248
First electron affinity of oxygen	−142
Second electron affinity of oxygen	+844
Lattice enthalpy of magnesium oxide	+3888

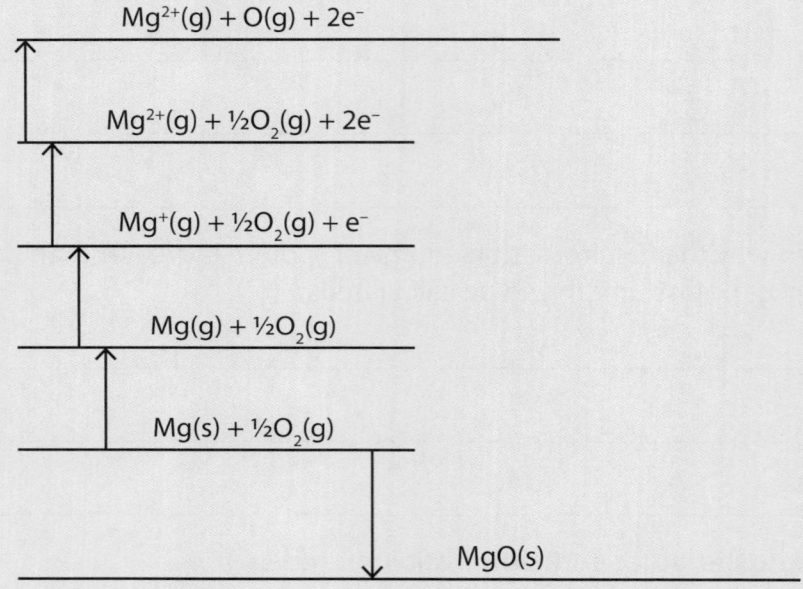

(a) Complete the Born-Haber cycle for magnesium oxide. [4]

(b) Calculate the enthalpy of formation of magnesium oxide.

_____

_____

_____

_____ [2]

4.2 Enthalpy, Entropy and Free Energy

Entropy and predicting entropy changes

1. Which one of the following reactions would show a decrease in entropy?

    A $Mg(s) + 2HCl(aq) \rightarrow MgCl_2(aq) + H_2(g)$

    B $N_2(g) + 3H_2(g) \rightarrow 2NH_3(g)$

    C $2NaHCO_3(s) \rightarrow Na_2CO_3(s) + CO_2(g) + H_2O(l)$

    D $2H_2O_2(aq) \rightarrow 2H_2O(l) + O_2(g)$ [1]

2. Which one of the following is true for the reaction below?

    $2Na(s) + 2H_2O(l) \rightarrow 2NaOH(aq) + H_2(g)$

    A ΔS is negative and ΔH is negative

    B ΔS is negative and ΔH is positive

    C ΔS is positive and ΔH is negative

    D ΔS is positive and ΔH is positive [1]

3. Which of the following are the units of entropy?

    A $kJ\ mol^{-1}$

    B $J\ mol^{-1}$

    C $J\ K^{-1}\ mol^{-1}$

    D $kJ\ K^{-1}$ [1]

Calculating ΔS, ΔH and ΔG

4. Some standard entropy values are given below:

    $S^\ominus / J\ K^{-1}\ mol^{-1}$ $H_2O(l) = 70$ $H_2O(g) = 189$

    Water boils at 373 K. Which one of the following is the enthalpy of vaporisation of water if ΔG = 0 when boiling occurs?

    A $+11.9\ kJ\ mol^{-1}$

    B $+25.9\ kJ\ mol^{-1}$

    C $+44.4\ kJ\ mol^{-1}$

    D $+96.6\ kJ\ mol^{-1}$ [1]

5 The reaction below is feasible at temperatures below 794.3 K.

NO(g) + ½O₂(g) → NO₂(g)   ΔH⦵ = −56 kJ mol⁻¹

(a) What is meant by **feasible**?

_____

_____ [1]

(b) The standard enthalpy of formation of NO₂(g) is + 34 kJ mol⁻¹. Calculate the standard enthalpy of formation of NO(g).

_____

_____

_____ [2]

(c) Calculate the value of ΔS⦵ for this reaction in J K⁻¹ mol⁻¹.

_____

_____

_____ [3]

(d) The standard entropy values for NO(g) and NO₂(g) are 208 and 240 J K⁻¹ mol⁻¹ respectively.

  (i) What is meant by the term **entropy**?

  _____

  _____ [1]

  (ii) Calculate the standard entropy of O₂(g).

  _____

  _____

  _____ [3]

4.2 ENTHALPY, ENTROPY AND FREE ENERGY

6 The equations for 3 reactions are given in the table below with some thermodynamic data.

	Reaction	ΔS⦵ / J K⁻¹ mol⁻¹	ΔH⦵ / kJ mol⁻¹
1	$CaCO_3(s) \rightarrow CaO(s) + CO_2(g)$	+160	+178
2	$2Mg(s) + O_2(g) \rightarrow 2MgO(s)$	−217	−1204
3	$H_2(g) + Cl_2(g) \rightarrow 2HCl(g)$	+20	−185

(a) What is the standard enthalpy of formation of magnesium oxide?

_____ [1]

(b) For reaction 1, calculate the temperature at which the reaction becomes feasible.

_____

_____

_____ [2]

(c) For reaction 2, the standard entropy values for Mg(s) and MgO(s) are 32.7 J K⁻¹ mol⁻¹ and 26.8 J K⁻¹ mol⁻¹. Calculate the standard entropy value for $O_2(g)$.

_____

_____ [2]

(d) For reaction 2, calculate the value of ΔG⦵ at 298 K and explain whether the reaction is feasible at this temperature.

_____

_____ [2]

(e) For reaction 3, explain why this reaction is feasible at any temperature.

_____

_____ [2]

7 The table below gives some data about the melting of magnesium.

Enthalpy of fusion (melting) of magnesium	+8.95 kJ mol^{-1}
Standard entropy of Mg(s)	32.7 J K^{-1} mol^{-1}
Standard entropy of Mg(l)	42.7 J K^{-1} mol^{-1}

(a) Calculate the entropy change for Mg(s) → Mg(l).

_____

_____ [1]

(b) Assuming the ΔG = 0 when melting occurs, calculate the melting point of magnesium in kelvin (K).

_____

_____

_____ [2]

8 Copper(II) nitrate decomposes on heating as shown by the equation below:

$$2Cu(NO_3)_2(s) \rightarrow 2CuO(s) + 4NO_2(g) + O_2(g)$$

The table below gives some data about this reaction.

	Cu(NO$_3$)$_2$(s)	CuO(s)	NO$_2$(g)	O$_2$(g)
Δ$_f$H$^⦵$ / kJ mol^{-1}	−302.9	−155.2	+33.8	0
S$^⦵$ / J K^{-1} mol^{-1}	193	43.5	240	205

(a) Explain why the enthalpy of formation of oxygen is zero.

_____

_____ [1]

(b) Calculate the value of ΔS$^⦵$ for this reaction.

_____

_____

_____ [2]

4.2 ENTHALPY, ENTROPY AND FREE ENERGY

(c) Calculate the value for ΔH⁻ for this reaction.

_____

_____

_____ [2]

(d) Calculate the value of ΔG⁻ at 500 K and explain whether the reaction is feasible at this temperature.

_____

_____

_____ [2]

(e) Calculate the minimum temperature required for this decomposition.

_____

_____

_____ [2]

9 Group II hydroxides decompose on heating to form oxides and release water vapour according to the general equation below.

$M(OH)_2(s) \rightarrow MO(s) + H_2O(g)$

Some information on these reactions is provided in the table below.

	Ba(OH)₂(s)	BaO(s)	Ca(OH)₂(s)	CaO(s)	H₂O(g)
Δ_fH⁻ / kJ mol⁻¹	to be calculated	−533.5	−986.1	−635.1	−242.0
S⁻ / J K⁻¹ mol⁻¹	99.7	70.4	83.4	39.7	189

(a) Suggest why the standard entropy of calcium oxide is lower than that of barium oxide.

_____

_____ [2]

(b) Barium hydroxide decomposes at temperatures of 1060 K and above.

$Ba(OH)_2(s) \rightarrow BaO(s) + H_2O(g)$

(i) Calculate the entropy change for this reaction.

_____ [2]

(ii) Calculate the enthalpy change for the reaction.

_____ [3]

(iii) Calculate the enthalpy of formation of barium hydroxide.

_____ [3]

(c) Calcium hydroxide decomposes according to the equation:

$Ca(OH)_2(s) \rightarrow CaO(s) + H_2O(g)$

Calculate the minimum temperature required for this decomposition.

_____ [6]

(d) The decomposition temperatures of magnesium hydroxide, strontium hydroxide and barium hydroxide are given in the table below. The trend is similar to that of the decomposition temperature of the Group II carbonates.

Group II hydroxide	Decomposition temperature / K
Mg(OH)$_2$(s)	529
Sr(OH)$_2$(s)	804
Ba(OH)$_2$(s)	1060

(i) Suggest why the decomposition temperature of the Group II hydroxides increases down the group.

_____

_____

_____

_____ [3]

(ii) State the trend in the solubility of the Group II hydroxides going down the group.

_____ [1]

10 Ammonia reacts with chlorine according to the equation below:

8NH$_3$(g) + 3Cl$_2$(g) → N$_2$(g) + 6NH$_4$Cl(s)

The table below gives the enthalpy of formation and entropy data for the reactants and products.

	NH$_3$(g)	Cl$_2$(g)	N$_2$(g)	NH$_4$Cl(s)
$\Delta_f H^\ominus$ / kJ mol^{-1}	−46.0	0	0	−315.5
$S^\ominus$ / J K^{-1} mol^{-1}	192.5	233.0	191.4	94.6

(a) Calculate the enthalpy change for this reaction.

_____

_____

_____ [2]

(b) Calculate the entropy change for this reaction.

_____

_____

_____ [2]

(c) Calculate the free energy change for this reaction at 1500 K and explain whether the reaction is feasible at that temperature.

_____

_____

_____ [2]

11 Methane reacts with steam to form carbon monoxide and hydrogen according to the equation:

$$CH_4(g) + H_2O(g) \rightarrow CO(g) + 3H_2(g)$$

The table below gives information about the reaction. $\Delta G = -10$ kJ mol^{-1} at 1000 K.

	$CH_4(g)$	$H_2O(g)$	$CO(g)$	$H_2(g)$
$\Delta_f H^\ominus$ / kJ mol^{-1}	−75	−242	−111	0
$S^\ominus$ / J K^{-1} mol^{-1}	186	189	198	to be calculated

(a) Calculate the enthalpy change (ΔH) for this reaction.

_____

_____

_____ [2]

(b) Write an expression for ΔG.

_____ [1]

(c) Use the expression for ΔG to calculate the entropy change for this reaction in J K^{-1} mol^{-1}.

_____

_____

_____

_____ [3]

(d) Use the calculated in (c) to determine the standard entropy of $H_2(g)$.

_____

_____

_____ [2]

(e) Calculate the minimum temperature at which this reaction becomes feasible.

_____

_____

_____ [2]

4.3 Rates of Reaction

Rate equations and associated graphs

1. The equation for a reaction is shown below with its rate equation.

   A + 2B → C + D

   rate = k [A]^2

   Which one of the following is **not** correct?

   A   The reaction is overall second order

   B   The units of the rate constant are mol^{-2} dm^6 s^{-1}

   C   The reaction is zero order with respect to B

   D   The rate of disappearance of B is twice the rate of disappearance of A   [1]

2. For the following reaction: W + 3X → Y + 2Z

   the rate equation is rate = k [W]^2 [X].

   Which one of the following graphs is correct for this reaction?

   A
   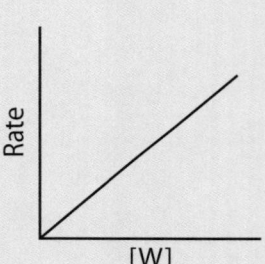

   B
   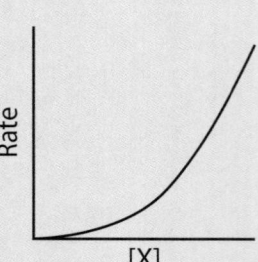

   C
   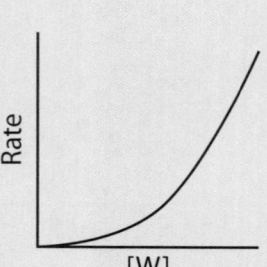

   D
   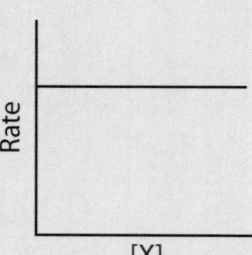

   [1]

4.3 RATES OF REACTION

3 L reacts with M according to the equation:

L + 2M → N

When the concentrations of L and M are both doubled, the rate increases by a factor of 8.

Which one of the following is a possible rate equation for the reaction?

A rate = k [L][M]

B rate = k [L]^2

C rate = k [L]^2[M]

D rate = k [M]^2 [1]

4 The reaction between P and Q occurs according to the equation below:

2P + 1½Q → R + 3S

The rate equation for the reaction is: rate = k[P]^2[Q]

The table below shows data for the initial rate of reaction for different initial concentrations of P and Q. The missing values will be calculated.

Experiment	[P] / mol dm^{-3}	[Q] / mol dm^{-3}	Rate of reaction / mol dm^{-3} s^{-1}
1	1.25 × 10^{-3}	1.75 × 10^{-3}	6.25 × 10^{-4}
2	to be calculated	1.75 × 10^{-3}	2.50 × 10^{-3}
3	2.00 × 10^{-3}	to be calculated	2.56 × 10^{-3}
4	2.50 × 10^{-3}	5.60 × 10^{-3}	to be calculated

(a) For experiment 2, calculate the concentration of P. Show your working out.

_____

_____

_____ [2]

(b) For experiment 3, calculate the concentration of Q. Show your working out.

_____

_____

_____ [2]

(c) For experiment 4, calculate the rate of reaction. Show your working out.

_____

_____

_____ [2]

(d) State the overall order of the reaction.

_____ [1]

(e) Using experiment 1, calculate a value for the rate constant, k, and state its units. Give your answer to 3 significant figures.

_____

_____

_____ [2]

Rate related to mechanism

5 For the reaction 2A + 2B → C + 3D, the reaction is second order with respect to A and zero order with respect to B.

(a) Explain what is meant by order of reaction.

_____

_____ [1]

(b) Write a rate equation for this reaction.

_____ [1]

(c) A proposed mechanism for the reaction is shown below.

Step 1:   2A → X + 2Y
Step 2:   X + B → C + D
Step 3:   2Y + B → 2D

(i) What is meant by the term **rate determining step**?

_____

_____ [1]

(ii) State and explain which step in this mechanism is the rate determining step.

_____

_____

_____

_____ [3]

6 Some information on the rate of hydrolysis of a halogenoalkane, C_4H_9Br, is given in the table below.

Experiment	[C_4H_9Br] / mol dm^{-3}	[OH$^-$] / mol dm^{-3}	Initial rate of reaction / mol dm^{-3} s^{-1}
1	0.0250	0.0250	1.20×10^{-3}
2	0.0450	0.0250	2.16×10^{-3}
3	0.0675	0.0750	9.72×10^{-3}

(a) Determine the order of reaction with respect to the C_4H_9Br and OH$^-$.

C_4H_9Br _____

OH$^-$ _____ [2]

(b) Write a rate equation for this reaction.

_____ [1]

(c) Calculate a value for the rate constant, k, using the results of experiment 1 and state its units.

_____

_____

_____ [2]

(d) Suggest how the value of the rate constant would change as temperature changes.

_____

_____ [1]

(e) Suggest a possible identity of this halogenoalkane and explain your reasoning.

_____

_____

_____

_____ [3]

Practical determination of rate and graphical analysis

7 Dinitrogen pentoxide, N$_2$O$_5$, decomposes to form nitrogen(IV) oxide and oxygen gas.

2N$_2$O$_5$ → 4NO$_2$ + O$_2$

The dinitrogen pentoxide and nitrogen(IV) oxide remain dissolved in the solvent used and oxygen gas is released.

(a) Describe how you would carry out an experiment to determine the initial rate of this reaction.

_____

_____

_____

_____ [4]

(b) The graph below shows how the concentration of N$_2$O$_5$ changes over time. Two tangents are drawn on the graph at 2.50 mol dm^{-3} (labelled A) and at 1.00 mol dm^3 (labelled B).

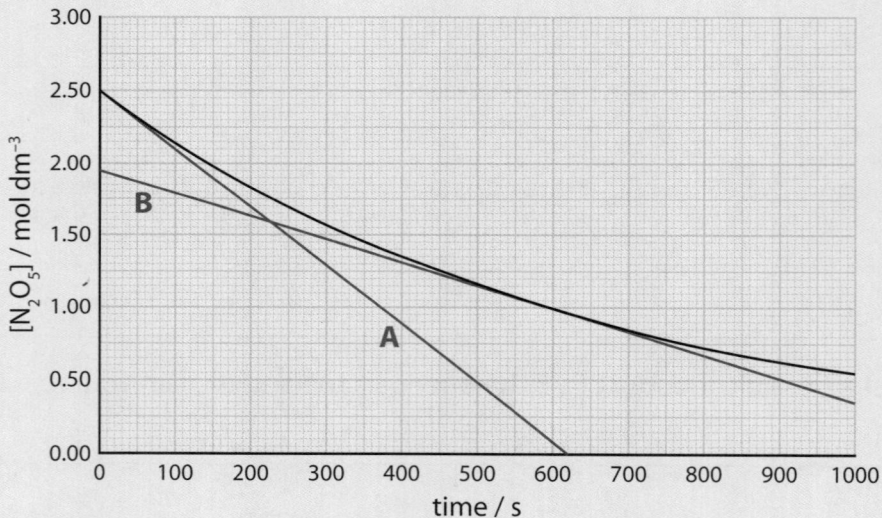

(i) Using the graph, determine the time at which the concentration of N$_2$O$_5$ was half of its initial value.

_____ [1]

(ii) Determine the gradient of the two tangents and use this to complete the table below.

[N₂O₅] / mol dm⁻³	Rate of reaction / mol dm⁻³ s⁻¹
2.50	
1.00	

[2]

(iii) Based on your answers to (ii) above, determine the order of reaction with respect to N₂O₅. Explain your answer.

_____

_____

_____ [2]

(iv) Write a rate equation for this reaction.

_____ [1]

(v) Calculate a value for the rate constant and state its units.

_____

_____ [2]

8 Iodine reacts with propanone in aqueous solution according to the equation:

$CH_3COCH_3(aq) + I_2(aq) \rightarrow CH_3COCH_2I(aq) + H^+(aq) + I^-(aq)$

Hydrogen ions catalyse the reaction.

(a) Describe how the order with respect to iodine could be determined experimentally using colorimetry.

_____

_____

_____

_____

_____

_____

_____

_____

_____

_____ [6]

(b) The graph below shows how the concentration of iodine changes against time.

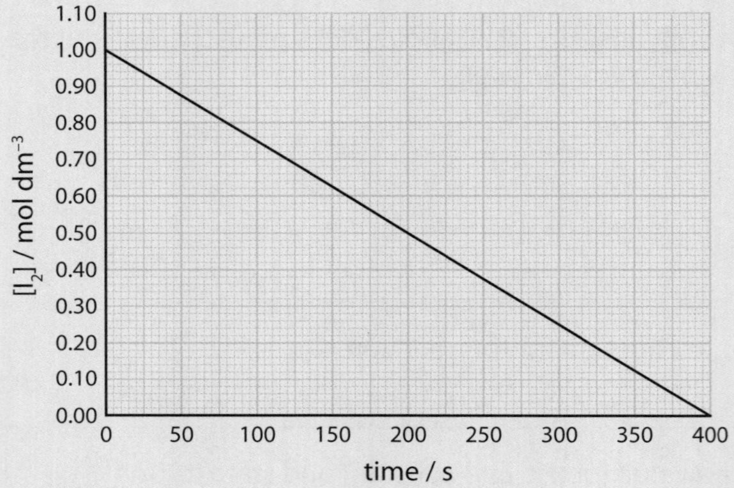

[Graph: $[I_2]$ / mol dm^{-3} vs time / s, showing a straight line from (0, 1.00) to (400, 0.00)]

(i) State the order with respect to iodine based on the graph and explain your answer.

_____

_____

_____ [2]

(ii) Calculate the rate of reaction during this experiment from the graph.

_____

_____ [1]

(iii) Sketch a graph of rate of reaction against concentration of iodine on the axes below.

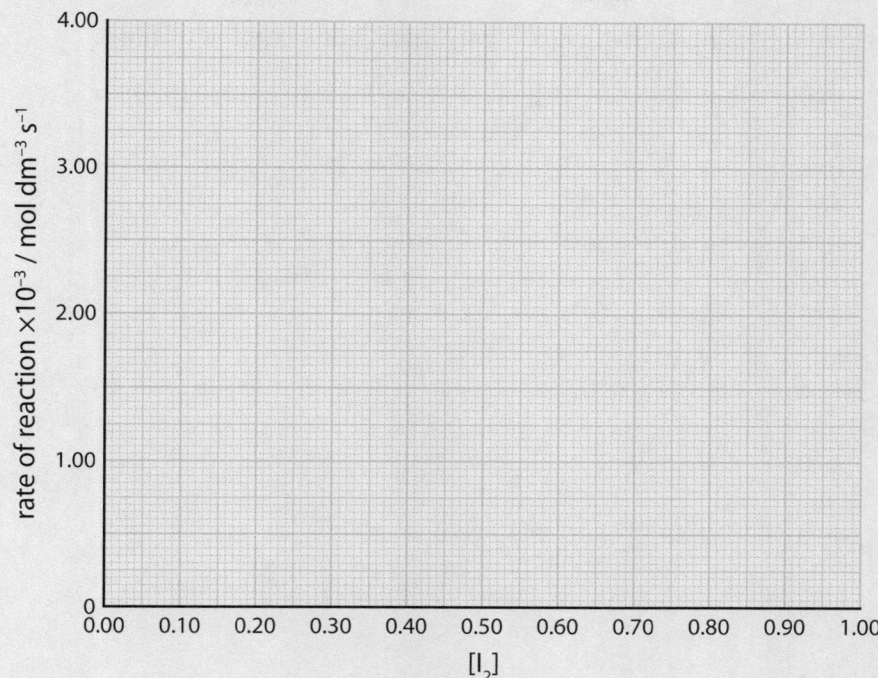

[2]

(c) The table below shows data obtained from experiments to determine the order of reaction with respect to propanone and hydrogen ions.

Experiment	[CH$_3$COCH$_3$] / mol dm^{-3}	[H$^+$] / mol dm^{-3}	[I$_2$] / mol dm^{-3}	Initial rate of reaction / mol dm^{-3} s^{-1}
1	0.150	0.150	1.00	2.35 × 10^{-4}
2	0.300	0.150	1.00	4.70 × 10^{-4}
3	0.300	0.300	1.00	9.40 × 10^{-4}

(i) Write the rate equation for this reaction, based on your answer to (b)(i) and using the table above.

_____ [1]

(ii) Calculate a value for the rate constant, k, using the results of experiment 1. State the units.

_____

_____

_____ [2]

A2 1: FURTHER PHYSICAL AND INORGANIC CHEMISTRY

9 A reacts with B according to the equation:

A + 2B → C + D

The graph below shows how the concentration of A changes with time.

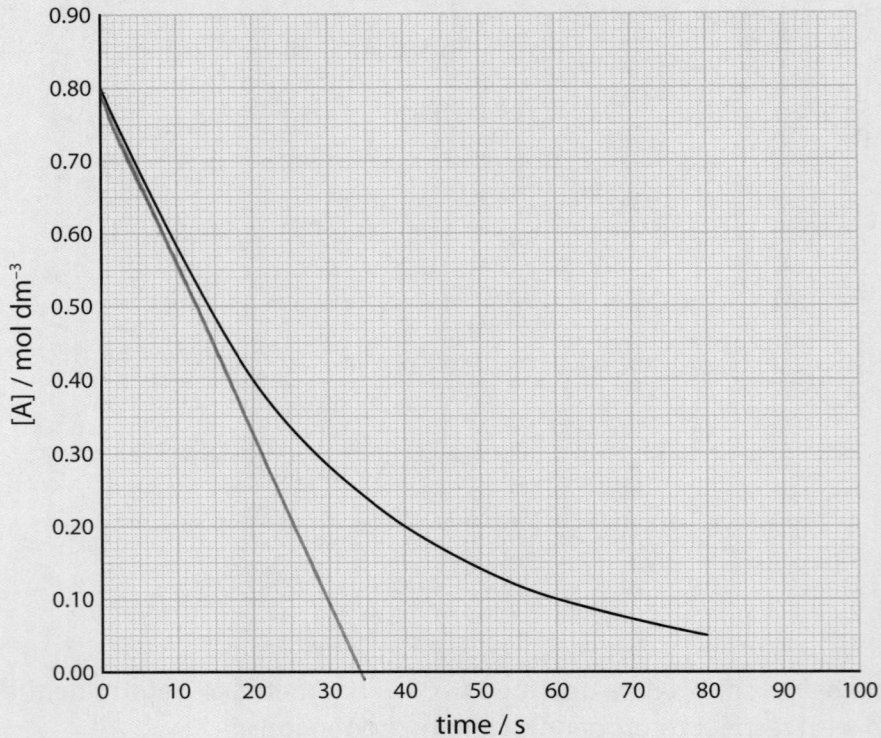

(a) Draw a tangent to the curve at t = 0 s. [1]

(b) Using the tangent, calculate the initial rate of the reaction. State the units.

_____

_____ [2]

(c) Use the graph to state the time at which show the time at which the concentration of iodine was at the shown values.

0.40 mol dm⁻³ _____

0.20 mol dm⁻³ _____ [1]

(d) Based on your answers to (c), state the order of reaction with respect to A.

_____ [1]

(e) The reaction is zero order with respect to B. Write a rate equation for the reaction.

_____ [1]

(f) Using your answers to (b) and (e), calculate a value for the rate constant, k, and state its units.

_____

_____

_____ [2]

(g) The mechanism for this reaction occurs in three steps:

Step 1:   A → X + Y

Step 2:   X + B → C

Step 3:   Y + B → D

Explain which step in the mechanism is the rate determining step.

_____

_____

_____ [1]

4.4 Equilibrium

1 For the equilibrium reaction:

$$CH_3CH_2OH + CH_3COOH \rightleftharpoons CH_3CH_2OOCCH_3 + H_2O$$

1.0 mole of CH_3CH_2OH and 1.0 mole of CH_3COOH were mixed and allowed to reach equilibrium at 25 °C. 0.35 moles of CH_3CH_2OH were present at equilibrium. Which of the following is the value of the equilibrium constant, K_c?

A   0.082

B   0.29

C   3.4

D   5.3 [1]

2 For the equilibrium:

$$SO_2(g) + \tfrac{1}{2}O_2(g) \rightleftharpoons SO_3(g)$$

$K_c = 2.40 \; mol^{-\frac{1}{2}} \; dm^{-\frac{3}{2}}$ at 450 °C.

What is the value of the equilibrium constant for the reaction below at the same temperature?

$$2SO_2(g) + O_2(g) \rightleftharpoons 2SO_3(g)$$

A   1.55

B   2.40

C   4.80

D   5.76 [1]

3 The equilibrium constant, K_c, for a reaction is:

$$K_c = \frac{[CO_2][H_2]^4}{[CH_4][H_2O]^2}$$

At 500 K, $K_c = 2.40 \times 10^2 \; mol^2 \; dm^{-6}$.

(a) Write the equation for the equilibrium reaction.

_____ [1]

(b) At 1000 K, $K_c = 5.45 \times 10^5 \; mol^2 \; dm^{-6}$. State and explain whether the forward reaction is exothermic or endothermic.

_____

_____ [1]

4.4 EQUILIBRIUM

(c) Another equilibrium involving methane and water vapour is shown below.

$$CH_4(g) + H_2O(g) \rightleftharpoons CO(g) + 3H_2(g)$$

2.45 moles of $CH_4(g)$ were mixed with 4.00 moles of $H_2O(g)$ in a container of volume 10.0 dm³. At equilibrium, 1.55 moles of $CO(g)$ were present.

(i) Write an expression for the equilibrium constant, K_c, for this reaction.

_____ [1]

(ii) Calculate the equilibrium concentrations, in mol dm⁻³, of all the reactants and products.

$CH_4(g)$ _____

$H_2O(g)$ _____

$CO(g)$ _____

$H_2(g)$ _____ [4]

(iii) Calculate a value for the equilibrium constant, K_c, and state its units. Give your answer to 3 significant figures.

_____

_____

_____ [3]

4 Hydrogen reacts with iodine according to the equilibrium below. The forward reaction is exothermic.

$$H_2(g) + I_2(g) \rightleftharpoons 2HI(g)$$

(a) A mixture of 1.90 moles of hydrogen and 1.90 moles of iodine was allowed to reach equilibrium at 710 K in a 2.00 dm³ container. The equilibrium mixture was found to contain 3.00 moles of hydrogen iodide.

(i) Write an expression for the equilibrium constant, K_c.

_____ [1]

(ii) Explain why K_c has no units.

_____

_____ [1]

(iii) Calculate a value for the equilibrium constant, K_c, at 710 K. Give your answer to 3 significant figures.

_____

_____

_____

_____

_____

_____ [3]

(b) State the effect, if any, of changing the temperature to 500 K on the position of equilibrium and the value of K_c.

_____

_____

_____

_____ [2]

(c) State and explain the effect, if any, of increasing the pressure on the position of equilibrium and the value of K_c.

_____

_____

_____

_____ [3]

4.4 EQUILIBRIUM

5 Ethanoic acid, CH_3COOH and pentene, C_5H_{10}, react to produce pentyl ethanoate in an inert solvent. A solution was prepared containing 0.020 moles of pentene and 0.010 moles of ethanoic acid in 600 cm³ of solution. At equilibrium there were 9.0×10^{-3} moles of pentyl ethanoate.

$$CH_3COOH + C_5H_{10} \rightleftharpoons CH_3COOC_5H_{11}$$

Calculate a value for the equilibrium constant, K_c, and state its units. Give your answer to an appropriate number of significant figures.

_____

_____

_____

_____

_____ [4]

6 For the reaction:

$$N_2(g) + 3H_2(g) \rightleftharpoons 2NH_3(g)$$

$K_c = 8.00$ mol⁻² dm⁶ at 1000 K.

(a) Calculate the value of K_c at 1000 K for the reaction below and state the units. Give your answer to 3 significant figures.

$$NH_3(g) \rightleftharpoons \tfrac{1}{2}N_2(g) + 1\tfrac{1}{2}H_2(g)$$

_____

_____

_____ [2]

(b) An unknown number of moles of nitrogen were mixed with 3.00 moles of hydrogen in a container of size 0.2 dm³. At 1000 K, equilibrium was established and 0.30 moles of hydrogen were present.

(i) Using x to represent the initial number of moles of nitrogen, calculate the equilibrium concentrations of nitrogen, hydrogen and ammonia in mol dm⁻³.

N₂ _____

H₂ _____

NH₃ _____ [3]

(ii) Using the value for the equilibrium constant at 1000 K, calculate the initial moles of nitrogen used.

_____ [4]

(iii) State and explain the effect, if any, of increasing the volume of the container on the position of equilibrium and the value of the equilibrium constant at 1000 K.

_____ [3]

7 For the reaction:

$$I_2(g) \rightleftharpoons 2I(g)$$

$K_c = 4.60 \times 10^{-3}$ mol dm^{-3} at 1200 K.

(a) Write an expression for the equilibrium constant, K_c.

_____ [1]

(b) 1.25×10^{-3} moles of $I_2(g)$ were placed in a container of volume 15.0 dm³ at 800 K. At equilibrium, 1.20×10^{-3} moles of I(g) were present. Calculate a value for the equilibrium constant, K_c, at 800 K. Give your answer to 3 significant figures.

_____ [4]

4.4 EQUILIBRIUM

(c) At 1200 K, another equilibrium was established in a different container. The equilibrium concentration of $I_2(g)$ was found to be 0.0458 mol dm^{-3}.

(i) Suggest how the equilibrium concentration of $I_2(g)$ could be measured.

_____

_____ [1]

(ii) Calculate the equilibrium concentration of $I(g)$ at 1200 K.

_____

_____

_____

_____ [3]

(d) State and explain using K_c values if the forward reaction is exothermic or endothermic.

_____

_____ [1]

8 Nitrogen and oxygen can react in the presence of ultraviolet light in the upper atmosphere to form nitrogen dioxide.

$$N_2(g) + 2O_2(g) \rightleftharpoons 2NO_2(g) \qquad \Delta H = +66 \text{ kJ mol}^{-1}$$

The conditions were mimicked in a 2.50 dm³ reaction vessel at 400 K. The graph below shows how the concentrations of nitrogen dioxide and oxygen change over a 7-minute period.

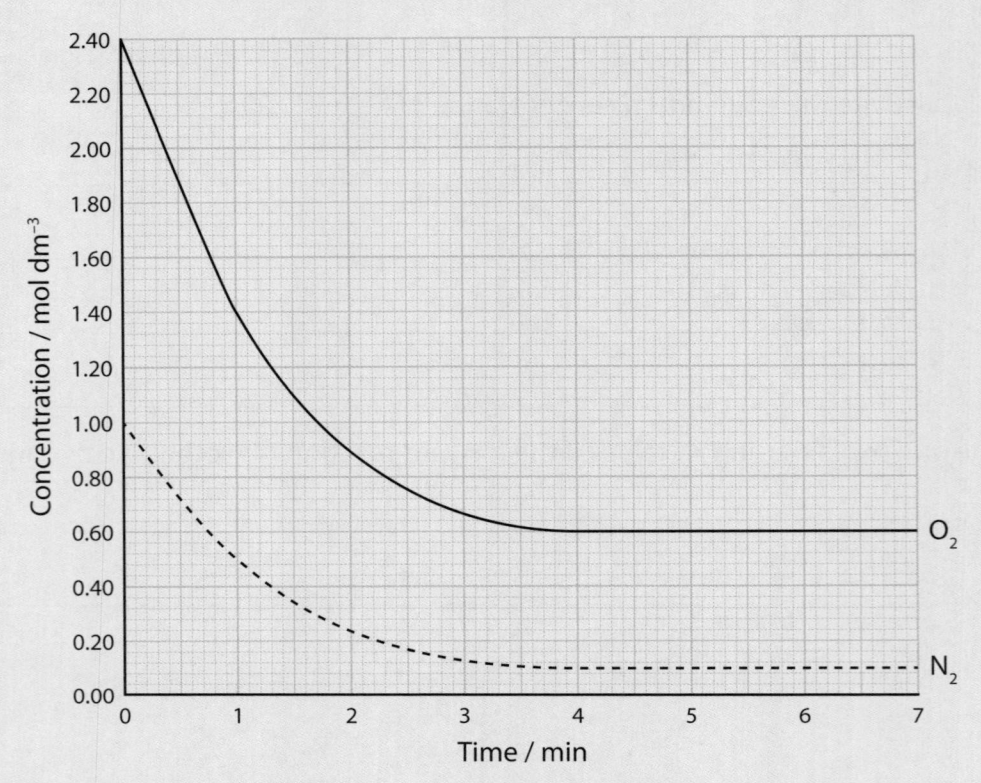

(a) State the time at which equilibrium is established. Explain your answer.

_____

_____ [2]

(b) State the initial and equilibrium concentrations of N_2 and O_2.

N₂   initial concentration   = _____ mol dm⁻³

O₂   initial concentration   = _____ mol dm⁻³

N₂   equilibrium concentration   = _____ mol dm⁻³

O₂   equilibrium concentration   = _____ mol dm⁻³   [4]

(c) Calculate the equilibrium concentration of NO_2.

_____

_____ [1]

(d) On the axes on the previous page sketch how the concentration of NO_2 changes over the 7 minutes. [2]

(e) Write an expression for the equilibrium constant, K_c, for this reaction. State the units.

_____

_____ [2]

(f) Calculate a value for the equilibrium constant, K_c, at 400 K.

_____

_____

_____ [2]

(g) State and explain whether you would expect K_c to change as temperature was increased.

_____

_____ [2]

A2 1: FURTHER PHYSICAL AND INORGANIC CHEMISTRY

9 (a) 2.50 moles of iodine and 2.50 moles of chlorine were mixed in a 4.00 dm³ vessel at 1000 K and allowed to react according to the equilibrium below.

$$I_2(g) + Cl_2(g) \rightleftharpoons 2ICl(g)$$

$K_c = 6.25 \times 10^{-2}$ at 1000 K.

(i) Write an expression for the equilibrium constant, K_c.

_____ [1]

(ii) Calculate the number of moles of I_2 present at equilibrium. Give your answer to 3 significant figures.

_____

_____

_____

_____ [4]

(c) In a different experiment, the graph below shows how the concentration of iodine changes over a 10 minute period at 300 K.

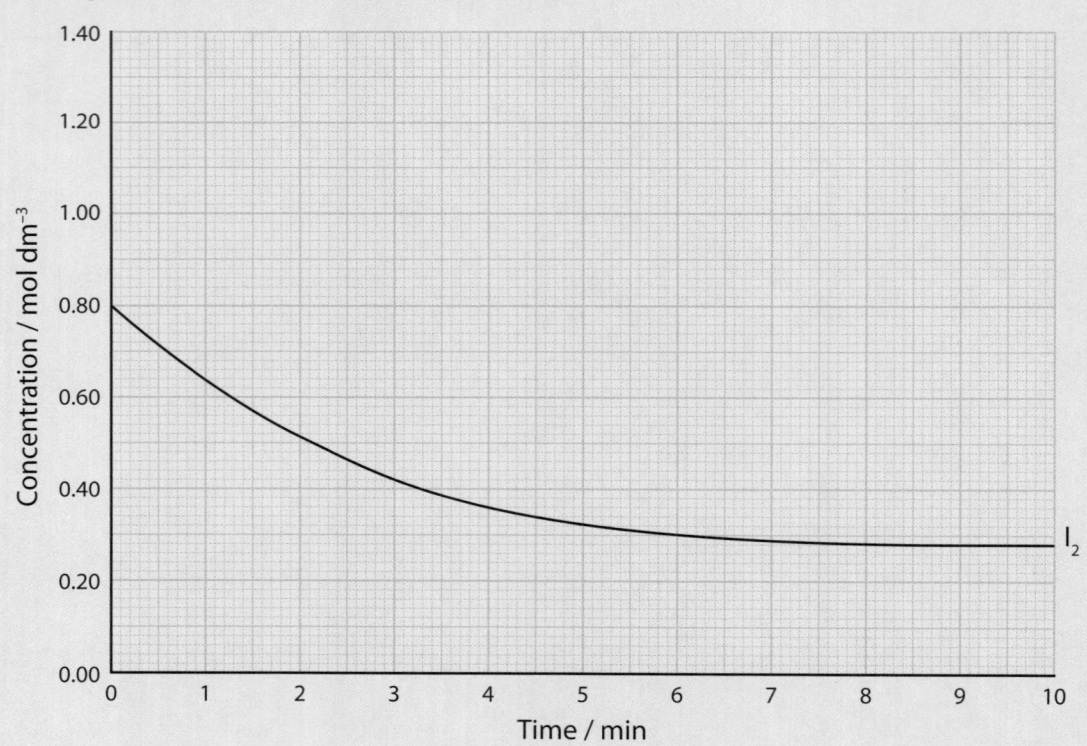

(i) In this experiment, the initial concentration of chlorine was 1.20 mol dm^{-3}. Sketch on the axes how the concentration of chlorine and iodine monochloride change over time. Label the graphs. [4]

(ii) Calculate the value for the equilibrium constant, K_c, for the reaction at 300 K. Give your answer to 3 significant figures.

_____

_____ [1]

(iii) State and explain whether the forward reaction is exothermic or endothermic.

_____

_____ [1]

4.5 Acid-Base Equilibria

Brønsted-Lowry acids and bases

1. Which one of the following is the conjugate acid of water?

    A H_2O

    B H_3O^+

    C OH^-

    D O^{2-} [1]

2. An acid-base equilibrium is set up between the two weak acids ethanoic acid, CH_3COOH, ($K_a = 1.75 \times 10^{-5}$ mol dm^{-3}) and chloroethanoic acid, $CH_2ClCOOH$, ($K_a = 1.38 \times 10^{-3}$ mol dm^{-3}).

    Complete the acid-base equilibrium below. Write **acid** or **base** in the boxes below all the species to show if they are acting as an acid or as base in this equilibrium.

    CH_3COOH + $CH_2ClCOOH$ ⇌   _____   +   _____

    [ ]   [ ]   [ ]   [ ]

    [2]

3. The hydrogenphosphate(V) ion, HPO_4^{2-}, can acid as a Brønsted-Lowry acid and base.

    (a) What is a Brønsted-Lowry acid?

    _____ [1]

    (b) Write an equation for the reaction of the hydrogenphosphate(V) ion acting as a Brønsted-Lowry base with water and name the ions formed.

    _____

    _____ [2]

    (c) Write an equation for the hydrogenphopshate(V) ion acting as a Brønsted-Lowry acid with water and name the ions formed.

    _____

    _____ [2]

4 A mixture of concentrated nitric acid and concentrated sulfuric acid is used to nitrate benzene and related compounds.

The reaction between nitric acid and sulfuric acid occurs in 3 stages.

Stage 1: $H_2SO_4 + HNO_3 \rightarrow HSO_4^- + H_2NO_3^+$

Stage 2: $H_2NO_3^+ \rightarrow NO_2^+ + H_2O$

Stage 3: $H_2SO_4 + H_2O \rightarrow H_3O^+ + HSO_4^-$

(a) Write an overall equation for this reaction.

_____ [1]

(b) In stage 1, identify the conjugate acid-base pairs.

_____
_____
_____
_____ [4]

(c) In stage 3, explain whether water is acting as a Brønsted-Lowry acid or base.

_____
_____ [1]

(d) Explain which of the two strong acids, nitric acid and sulfuric acid, is the stronger acid.

_____
_____ [1]

5 Which of the following does not show an acid-base reaction?

A $NaOH + HCl \rightarrow NaCl + H_2O$

B $H_2O + CH_3CH_2NH_2 \rightarrow OH^- + CH_3CH_2NH_3^+$

C $Na_2O + H_2O \rightarrow 2NaOH$

D $H_2 + Cl_2 \rightarrow 2HCl$ [1]

pH calculations

6 Which one of the following is **not** the correct pH of the solutions shown at 25 °C?

$K_w = 1.00 \times 10^{-14}$ mol² dm⁻⁶ at 25 °C

	Solution	pH
A	0.250 mol dm⁻³ hydrochloric acid	0.60
B	0.128 mol dm⁻³ sodium hydroxide solution	13.11
C	0.150 mol dm⁻³ sulfuric acid	0.52
D	0.178 mol dm⁻³ barium hydroxide solution	13.25

[1]

7 The table below shows values of K_w at different temperatures with some pH values.

Temperature / °C	K_w / mol² dm⁻⁶	pH
0	0.11×10^{-14}	7.48
25	1.0×10^{-14}	7.00
40	2.92×10^{-14}	to be calculated
100	to be calculated	6.15

(a) Write an expression for K_w.

_____ [1]

(b) Calculate the pH of water at 40 °C. Give your answer to 2 decimal places.

_____

_____ [2]

(c) Calculate the value of K_w at 100 °C. Give your answer to 3 significant figures.

_____

_____ [2]

4.5 ACID-BASE EQUILIBRIA

8 The table below shows dilutions of some strong acids. Complete the table. Give concentrations to 3 significant figures and pH values to 2 decimal places. Use the space below the table for working out. [6]

Strong acid	Concentration of original strong acid / mol dm^{-3}	Volume of strong acid used / cm^3	Volume of water added / cm^3	Concentration of diluted acid / mol dm^{-3}	pH of diluted acid
HCl	0.450	150	350		
HNO$_3$	0.597	100			1.40
H$_2$SO$_4$	0.250	25			1.38

_____

_____

_____

_____

_____

_____

9 Which of the following is the pH of a mixture of 20.0 cm^3 of 1.45 mol dm^{-3} sulfuric acid and 45.0 cm^3 of 2.40 mol dm^{-3} sodium hydroxide solution at 25 °C? $K_w = 1.00 \times 10^{-14}$ mol^2 dm^{-6} at 25 °C.

A 0.11

B 1.10

C 12.70

D 13.89 [1]

10 What volume of deionised water would be required to be added to 20 cm^3 of hydrochloric acid to change to pH from 1.75 to 2.45?

A 8 cm^3

B 20 cm^3

C 80 cm^3

D 100 cm^3 [1]

A2 1: FURTHER PHYSICAL AND INORGANIC CHEMISTRY

11 Barium oxide reacts with water to form a solution of barium hydroxide.

   (a) Write an equation for the reaction of barium oxide with water.

   _____ [1]

   (b) 3.15 g of barium oxide were added to water and the solution volume made up to 250 cm³ at 25 °C. Calculate the pH of the solution formed. Give your answer to 2 decimal places. $K_w = 1.00 \times 10^{-14}$ mol² dm⁻⁶ at 25 °C.

   _____

   _____

   _____

   _____

   _____ [5]

   (c) An excess of magnesium oxide was reacted with water in the same way as in (b) and the solution made up to 250 cm³. Explain why the pH of the solution formed here is less than the pH of the solution formed in (b).

   _____

   _____

   _____ [1]

12 A saturated solution of calcium hydroxide at 25 °C contains 168 mg dm⁻³.

   $K_w = 1.00 \times 10^{-14}$ mol² dm⁻⁶ at 25 °C. Give all pH values to 2 decimal places.

   (a) Calculate the pH of a saturated solution of calcium hydroxide at 25 °C. Give your answer to 2 decimal places.

   _____

   _____

   _____

   _____ [4]

(b) 25.0 cm³ of the saturated solution of calcium hydroxide were mixed with 5.0 cm³ of 0.0125 mol dm⁻³ hydrochloric acid. Calculate the pH of the resulting solution at 25 °C. Give your answer to 2 decimal places.

_____

_____

_____

_____

_____ [5]

(c) An additional 10.0 cm³ of 0.0125 mol dm⁻³ hydrochloric acid were added to the solution in (b). Calculate the pH of the resulting solution at 25 °C. Give your answer to 2 decimal places.

_____

_____

_____

_____

_____ [5]

13 The structure of citric acid is shown below. Citric acid is a weak tribasic acid.

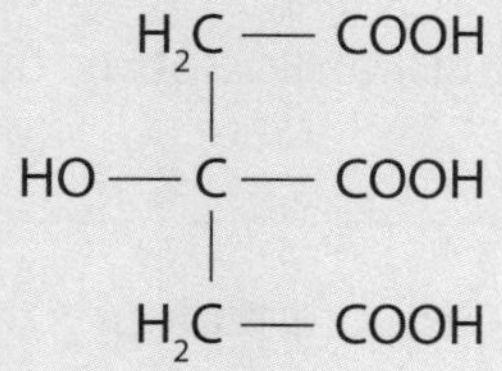

$$\begin{array}{c} H_2C-COOH \\ | \\ HO-C-COOH \\ | \\ H_2C-COOH \end{array}$$

(a) Write the molecular formula of citric acid.

_____ [1]

(b) The table below gives some information about citric acid.

$pK_a 1$	3.08
$pK_a 2$	4.74
$pK_a 3$	5.40

(i) Write equations for the three acid dissociations of citric acid.

_____

_____

_____ [2]

(ii) Calculate the pH of a solution of citric acid if 355 mg of solid citric acid were dissolved in 125 cm³ of deionised water. Use the first pK_a value only.

_____

_____

_____

_____

_____ [4]

14 (a) A solution of propanoic acid, CH_3CH_2COOH, of concentration 6.45×10^{-2} mol dm⁻³ has a pH of 3.03 at 25 °C.

(i) Write an equation for the acid dissociation of propanoic acid.

_____ [1]

(ii) Write an expression for the K_a for propanoic acid.

_____

_____ [1]

(iii) Calculate a value for K_a of the propanoic acid at 25 °C.

_____

_____

_____ [2]

(b) Another weak acid, butanoic acid, $CH_3CH_2CH_2COOH$, may be neutralised by sodium hydroxide. The K_a for butanoic acid is 1.51×10^{-5}.

(i) Write an equation for the reaction between butanoic acid and sodium hydroxide.

_____ [1]

(ii) Calculate the pH of a solution of butanoic acid of concentration 8.55×10^{-3} mol dm^{-3}. Give your answer to 2 decimal places.

_____ [2]

(iii) Calculate the volume of 1.20×10^{-2} mol dm^{-3} sodium hydroxide solution required to neutralise 25.0 cm^3 of 8.55×10^{-3} mol dm^{-3} butanoic acid. Give your answer to an appropriate number of significant figures.

_____ [2]

(iv) Calculate the pH when 25.0 cm^3 of the 8.55×10^{-3} mol dm^{-3} butanoic acid is mixed with 125 cm^3 of deionised water. Give your answer to 2 decimal places.

_____ [4]

15 A solution of methanoic acid, HCOOH, (pK_a = 3.75) has a pH of 2.77. In a titration, 25.0 cm^3 of this acid were neutralised by 16.0 cm^3 of 0.0258 mol dm^{-3} sodium hydroxide solution.

(a) Calculate the concentration, in mol dm^{-3}, of H$^+$ in the solution using the pH. Give your answer to 3 significant figures.

_____ [1]

A2 1: FURTHER PHYSICAL AND INORGANIC CHEMISTRY

(b) (i) Write an equation for the reaction of methanoic acid with sodium hydroxide.

_____ [1]

(ii) Calculate the concentration, in mol dm^{-3}, of H$^+$ ions based on the results of the titration. Give your answer to 3 significant figures.

_____

_____

_____ [2]

(iii) Explain why there is a difference in the values calculated in (a) and (b)(ii).

_____

_____

_____ [2]

Buffers

16 What is the pH of the buffer formed when 30.0 cm^3 of 0.1 M ethanoic acid (K_a = 1.75 × 10^{-5} mol dm^{-3}) are mixed with 15.0 cm^3 of 0.1 M sodium hydroxide?

A   2.88

B   4.02

C   4.76

D   5.05   [1]

17 Sodium lactate is an acidity regulator which is added to food. A buffer is formed when solid sodium lactate is added to a solution of lactic acid. The table below gives the structural formulae of lactic acid and sodium lactate.

Substance	Structural formula
lactic acid	H−C(H)(H)−C(H)(OH)−C(=O)(OH)
sodium lactate	H−C(H)(H)−C(H)(OH)−C(=O)(O$^-$Na$^+$)

54

4.5 ACID-BASE EQUILIBRIA

(a) What is meant by a buffer?

_____

_____ [1]

(b) Explain how a mixture of sodium lactate and lactic acid acts as a buffer when a small amount of hydrochloric acid is added.

_____

_____

_____ [2]

(c) Explain how a mixture of sodium lactate and lactic acid acts as a buffer when a small amount of sodium hydroxide is added.

_____

_____

_____ [2]

(d) Calculate the pH of the buffer formed when 216 mg of sodium lactate were added to 50.0 cm³ of 0.0125 mol dm⁻³ lactic acid. The K_a of lactic acid is 1.38×10^{-4} mol dm⁻³.

_____

_____

_____

_____

_____ [4]

A2 1: FURTHER PHYSICAL AND INORGANIC CHEMISTRY

18 Benzoic acid, C_6H_5COOH, is a weak acid with a pK_a of 4.19.

A buffer solution is formed when 35.0 cm³ of 0.250 mol dm⁻³ benzoic acid reacts with 15.0 cm³ of 0.250 mol dm⁻³ sodium hydroxide solution.

(a) Calculate the pH of the buffer solution formed. Give your answer to 2 decimal places.

[5]

(b) When acid is added to a buffer, the amount of the weak acid increases and the amount of the salt decreases by the same amount, as the salt reacts with the H⁺ ions to remove them.

$$C_6H_5COO^- + H^+ \rightarrow C_6H_5COOH$$
benzoate ions   benzoic acid

5.00 cm³ of hydrochloric acid of concentration 0.0550 mol dm⁻³ were added to the buffer.

(i) Calculate the number of moles of H⁺ ions added to the buffer.

[1]

(ii) Calculate the number of moles of benzoic acid present in the solution after the addition of the acid.

[1]

(iii) Calculate the number of moles of benzoate ions present in the solution after the addition of the acid.

[1]

(iv) Calculate the new pH of the buffer using the answers to (b)(ii) and (b)(iii).

_____

_____

_____

_____ [2]

(c) Explain using equations how this solution is able to act as a buffer when small amounts of alkali are added.

_____

_____

_____

_____ [2]

19 A buffer was prepared by mixing the following:

25.0 cm³ of 0.150 mol dm⁻³ nitrous acid (HNO_2)

15.0 cm³ of 0.100 mol dm⁻³ sodium hydroxide solution

The pK$_a$ for nitrous acid is 3.40.

(a) Explain how the buffer would maintain an approximate pH if a small sample of hydrochloric acid was added. Use an equation in your answer.

_____

_____

_____ [2]

(b) Calculate the pH of the buffer formed. Give your answer to 2 decimal places.

_____

_____

_____

_____

_____ [5]

(c) Calculate the volume of sodium hydroxide which would be required to form a buffer with pH 3.40 using the same concentrations and same volume of the acid.

_____

_____

_____

_____

_____ [3]

20 Ammonia and ammonium chloride form a buffer when they are mixed together in solution. The K_a for the equilibrium below is 5.62×10^{-10}.

$$NH_4^+ \rightleftharpoons NH_3 + H^+$$

(a) Write a K_a expression for this equilibrium.

_____ [1]

(b) The buffer may be formed by mixing excess ammonia solution with hydrochloric acid. Ammonia and hydrochloric acid react to leave a solution containing ammonia and ammonium ions.

  (i) Write an equation for the reaction of ammonia with hydrochloric acid.

_____ [1]

  (ii) Calculate the pH of the buffer formed when 50.0 cm³ of 0.1 M ammonia solution is mixed with 25.0 cm³ of 0.12 M hydrochloric acid. Give your answer to 2 decimal places.

_____

_____

_____

_____

_____ [5]

(c) Explain, using the equilibrium above, how the buffer maintains the pH when a small quantity of acid is added.

_____

_____

_____ [2]

Titration curves and salts

21 Which one of the following would have the lowest pH?

A   0.1 M ammonium chloride solution

B   0.1 M ammonium ethanoate solution

C   0.1 M potassium chloride solution

D   0.1 M potassium ethanoate solution [1]

22 The pK_a of ethanoic acid is 4.75 and the pK_a of chloroethanoic acid is 2.85. Which one of the following would have the highest pH?

A   0.25 mol dm^{-3} ammonium chloroethanoate

B   0.25 mol dm^{-3} ammonium ethanoate

C   0.25 mol dm^{-3} sodium chloroethanoate

D   0.25 mol dm^{-3} sodium ethanoate [1]

23 Hydrochloric acid reacts with sodium hydroxide.

(a) Write an equation for the reaction between hydrochloric acid and sodium hydroxide.

_____ [1]

(b) (i) Calculate the pH of a solution of 0.270 mol dm^{-3} hydrochloric acid. Give your answer to 2 decimal places.

_____

_____ [2]

(ii) Calculate the volume, in cm^3, of 0.320 mol dm^{-3} sodium hydroxide solution which is required to react with 25.0 cm^3 of 0.270 mol dm^{-3} hydrochloric acid. Give your answer to an appropriate number of significant figures.

_____

_____

_____ [3]

(iii) Sketch a titration curve on the axes below for the titration when 0.320 mol dm^{-3} sodium hydroxide solution is added to 25.0 cm^3 of 0.270 mol dm^{-3} hydrochloric acid. [3]

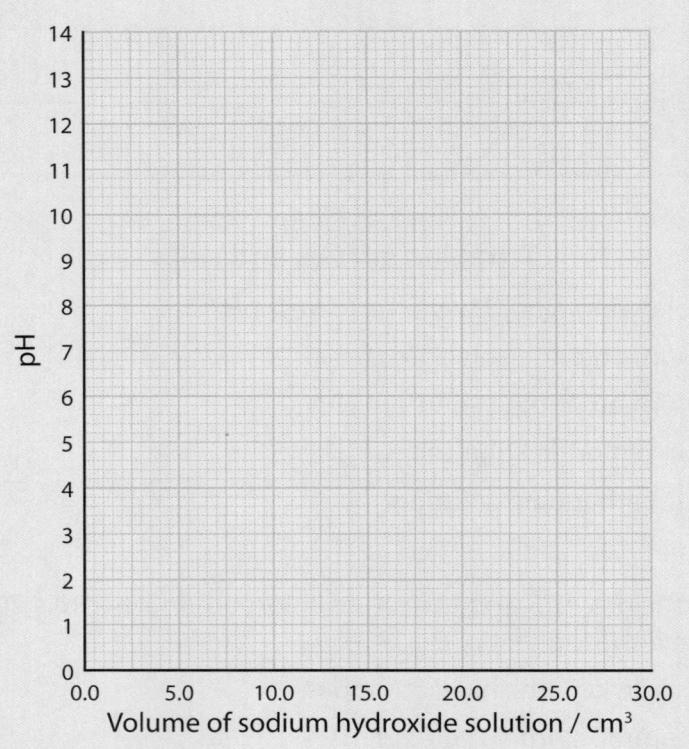

24 25.0 cm^3 of 0.145 mol dm^{-3} sulfuric acid were placed in a conical flask and titrated against 0.500 mol dm^{-3} potassium hydroxide solution. The equation for the reaction is:

   2KOH(aq) + H$_2$SO$_4$(aq) → K$_2$SO$_4$(aq) + 2H$_2$O(l)

Give all pH values to 2 decimal places.

K$_w$ = 1.00 × 10^{-14} mol^2 dm^{-6} at 25 °C.

(a) Calculate the pH of 0.145 mol dm^{-3} sulfuric acid.

_____

_____ [2]

(b) Calculate the initial pH of the potassium hydroxide solution.

_____

_____ [2]

(c) Calculate the volume of potassium hydroxide required for neutralisation. Give your answer to an appropriate number of significant figures.

_____

_____

_____ [3]

(d) Calculate the pH when 14.0 cm³ of potassium hydroxide solution had been added to the sulfuric acid.

_____

_____

_____

_____ [5]

(e) Calculate the pH when 15.0 cm³ of potassium hydroxide solution had been added to the sulfuric acid.

_____

_____

_____

_____ [5]

(f) On the axes below, sketch a titration curve to show the change in pH during this titration. [3]

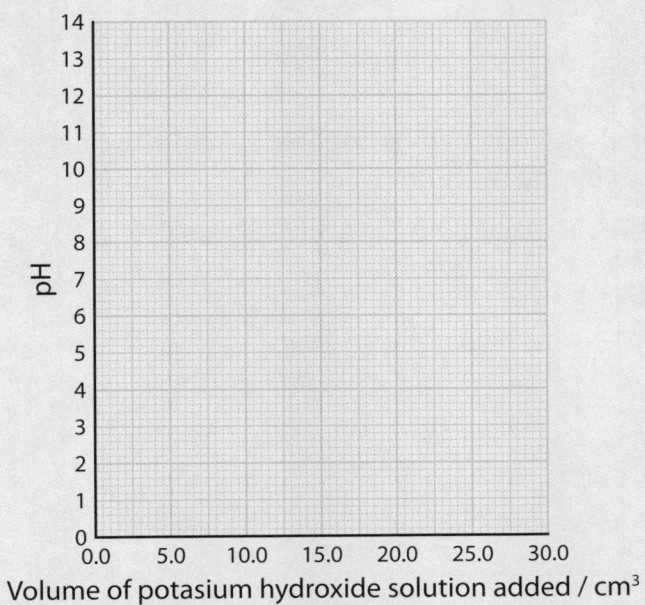

(g) Suggest a suitable indicator for this titration and explain your choice.

_____

_____

_____ [2]

25 The table below shows the pK_a values of some indicators. The pK_a of an indicator is the pH at which the indicator changes colour.

	Indicator	pKa
A	cresol red	1.0
B	bromophenol blue	4.1
C	phenolphthalein	9.5
D	thymolphthalein	9.2

Which indicator would be suitable for the titration of 0.100 mol dm^{-3} hydrochloric acid with ammonia solution? [1]

26 4-chlorobenzoic acid, ClC$_6$H$_4$COOH, is a weak organic acid. Its structure is shown below.

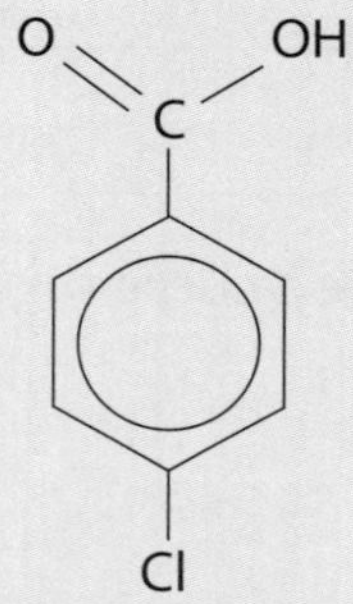

A 25.0 cm^3 sample of a solution of 4-chlorobenzoic acid was titrated with sodium hydroxide solution. The equation for the reaction is:

ClC$_6$H$_4$COOH(aq) + NaOH(aq) → ClC$_6$H$_4$COONa(aq) + H$_2$O(l)

A pH curve for the titration is shown below.

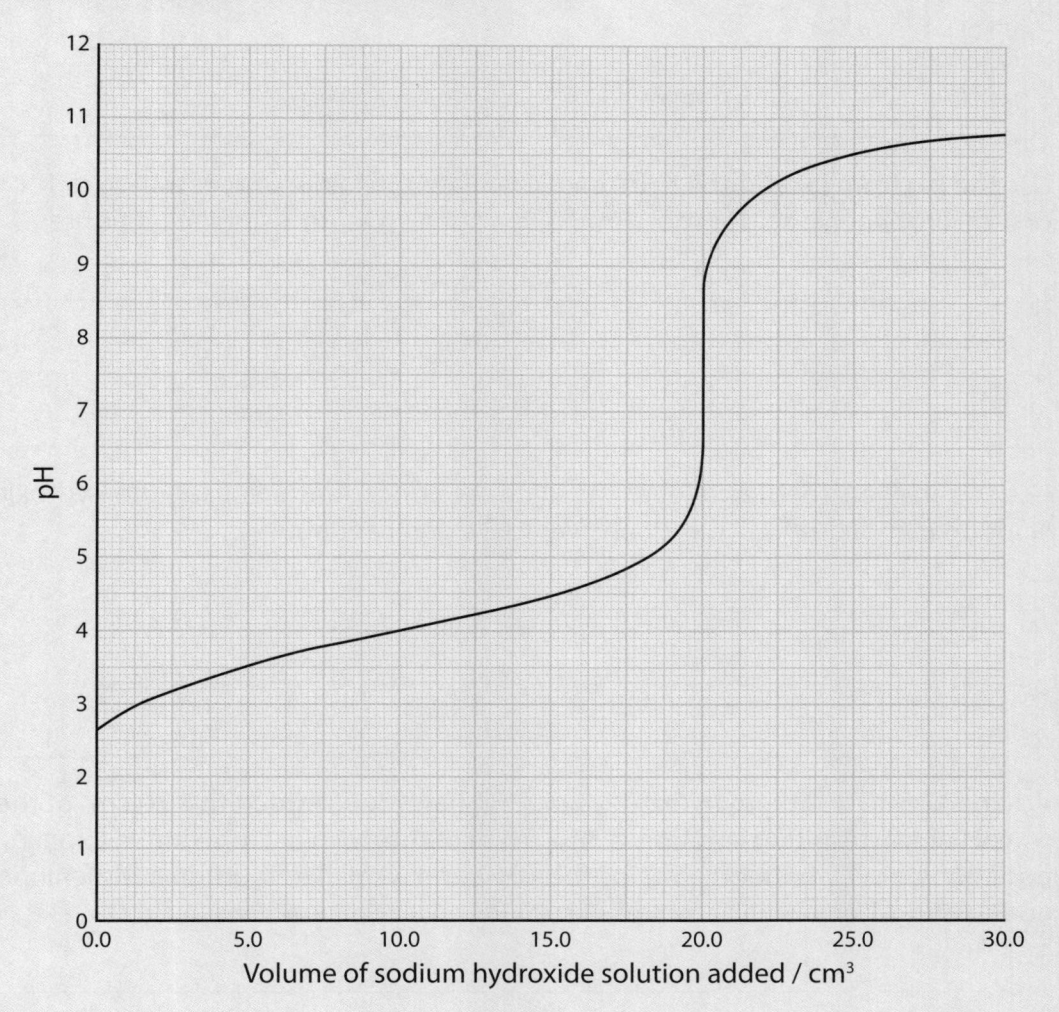

(a) From the graph state the initial pH of the solution of 4-chlorobenzoic acid.

_____ [1]

(b) From the graph, state the volume of sodium hydroxide solution required to neutralise the acid.

_____ [1]

(c) At half-neutralisation (when half the volume of alkali required to neutralise the acid has been added), the pH = pK_a. Use the graph to calculate the pK_a and K_a of 4-chlorobenzoic acid.

_____

_____

_____ [2]

(d) Write a K_a expression for 4-chlorobenzoic acid.

_____ [1]

(e) Use the pH from (a) and the K_a in (c) to calculate the initial concentration of the solution of 4-chlorobenzoic acid in mol dm^{-3}. Give your answer to 3 significant figures.

_____

_____

_____ [3]

(f) Use the answers to (b) and (e) to calculate the concentration of sodium hydroxide solution in mol dm^{-3}. Give your answer to 3 significant figures.

_____

_____

_____ [2]

(g) When 5.0 cm^3 of sodium hydroxide solution have been added to 25.0 cm^3 of the acid, the pH of the resulting solution is 3.52. This solution can act as a buffer. Explain, using equations, how a buffer maintains this pH when a small amount of hydrochloric acid is added and when a small amount of sodium hydroxide is added.

_____

_____

_____

_____ [4]

4.5 ACID-BASE EQUILIBRIA

27 The structure of the indicator methyl yellow is shown below. It is red below pH 2.9 and yellow above pH 4.0. It was added to butter to make it appear more yellow before concerns about its safety stopped it being used.

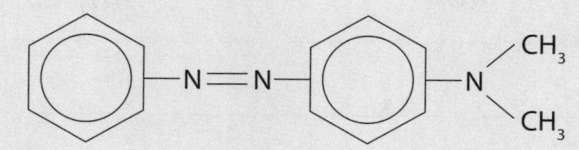

(a) On the molecule above indicate any positions where the indicator could form a co-ordinate bond with an H⁺ ion. [1]

(b) Details of four titrations labelled A to D are given below.

A   0.1 mol dm⁻³ ethanoic acid against 0.1 mol dm⁻³ ammonia solution

B   0.1 mol dm⁻³ ethanoic acid against 0.1 mol dm⁻³ sodium hydroxide solution

C   0.1 mol dm⁻³ nitric acid against 0.1 mol dm⁻³ ammonia solution.

D   0.1 mol dm⁻³ nitric acid against 0.1 mol dm⁻³ sodium hydroxide solution

(i) Write equations for the reactions which occur in each of the titrations.

A _____

B _____

C _____

D _____ [4]

(ii) In which titration(s) could methyl yellow be used? Explain your answer.

_____
_____
_____ [2]

(iii) The salt which is formed from titration B above is sodium ethanoate. A 0.1 mol dm⁻³ solution of sodium ethanoate has a pH 8.88. Explain, using an equation, why a solution of sodium ethanoate has an alkaline pH.

_____
_____
_____ [3]

(iv) The salt formed in titration C is ammonium nitrate. A 0.1 mol dm⁻³ solution of ammonium nitrate has a pH of 5.13. Explain, using an equation, why a solution of ammonium nitrate has an acidic pH.

_____

_____

_____ [3]

(v) A solution of the salt formed in D has a pH of 7.00. Explain why it forms a neutral solution.

_____

_____

_____ [2]

4.6 Isomerism

Optical isomers

1. Define the term **structural isomers**.

   _____

   _____ [2]

2. Draw and name two structural isomers of C_3H_6O which belong to different families of compounds.

   _____ [2]

3. (a) Draw and name two structural isomers of $C_3H_6O_2$ which belong to the same homologous series.

   _____ [2]

   (b) Draw and name a structural isomer of $C_3H_6O_2$ which belongs to a different homologous series than the isomers draw in 3(a).

   _____ [1]

4 Define the term **asymmetric (chiral) centre**.

_____ [1]

5 State if the molecules below have chiral centres.

(a) CH₃CH₂CH₂CH₂OH _____ [1]

(b) CH₃CH₂CH₂C(OH)₂CH₂CH₃ _____ [1]

(c) CH₃CHClCOOH _____ [1]

(d) CH₃CH(CH₃)CO₂H _____ [1]

(e) CH₃CH(OH)CO₂CH₃ _____ [1]

(f) CH₃CH(OH)CH₂C(OH)₂CH₂CH₂ _____ [1]

(g) CH₃CH₂CH₂CHClCH₃ _____ [1]

6 Which one of the following does not contain an asymmetric centre?

A   CH₃CH(OH)CH₂CH₃

B   CH₃CHClCH₂CH₃

C   CH₃CH(NH₂)CH₂CH₃

D   CH₃CH(CH₃)CH₂CH₃   [1]

7 Which one of the following displays optical isomerism?

A   CH₃CHOHCOOH

B   CH₂OHCH₂COOH

C   CH₂OHCOOCH₃

D   CH₂OHCH₂CHO   [1]

8 The structure of malic acid is shown below. It has an asymmetric centre and can exist as optical isomers.

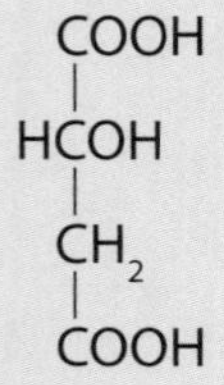

COOH
|
HCOH
|
CH₂
|
COOH

(a) Circle the asymmetric centre on the structure of malic acid.   [1]

(b) Explain what is meant by the term **optical isomers**.

_____

_____ [1]

(c) Name the two functional groups in malic acid.

_____ [1]

9 2-chlorobutane has structural and optical isomers.

   (a) Draw the 3D representations of the optical isomers of 2-chlorobutane.

[2]

   (b) Describe how you could distinguish between separate samples of the two optical isomers of 2-chlorobutane.

   _____

   _____

   _____ [2]

   (c) Draw and name a structural isomer of 2-chlorobutane and explain why it does not have optical isomers.

   _____ [2]

A2 1: FURTHER PHYSICAL AND INORGANIC CHEMISTRY

10 Lactic acid, CH₃CH(OH)COOH, is a compound which builds up in muscles during exercise.

   (a) Explain why lactic acid shows optical isomerism.

   _____ [1]

   (b) Draw the 3D structures of the two optical isomers of lactic acid. Give the bond angle around the central atom.

       Bond angle = _____ [3]

   (c) Give the structure of the organic compound which is produced when lactic acid is warmed with acidified potassium dichromate(VI) solution.

   [1]

   (d) State the IUPAC name for lactic acid.

   _____ [1]

   (e) Explain how you could distinguish between a racemate of lactic acid and one of the enantiomers of lactic acid.

   _____

   _____

   _____

   _____ [4]

4.6 ISOMERISM

11 Which one of the following can exhibit both geometrical and optical isomerism?

A   (CH$_3$)$_2$C=CHCH(CH$_3$)CH$_2$CH$_3$

B   CH$_3$CH$_2$CH=CHCH$_2$CH(CH$_3$)Cl

C   (CH$_3$)$_2$C=C(CH$_2$CH$_3$)$_2$

D   CH$_3$CH$_2$CH(CH$_3$)CH$_3$ [1]

12 Ethanal reacts with cold dilute potassium carbonate solution to yield 3-hydroxybutanal CH$_3$CH(OH)CH$_2$CHO which is optically active and was once used in medicinal drugs as a sedative.

(a) Explain the meaning of the term **optically active**.

_____

_____ [2]

(b) Draw the 3D structures for the two optical isomers of 3-hydroxybutanal.

[2]

(c) 3-hydroxybutanal in a sedative was used as a racemic mixture. Define this term and explain why a racemic mixture is optically inactive.

_____

_____ [2]

(d) Suggest how the drug action may be determined by the stereochemistry of the drug.

_____

_____ [1]

A2 1: FURTHER PHYSICAL AND INORGANIC CHEMISTRY

13 There are a number of structural isomers of C_4H_9OH. However, only one has an asymmetric centre and can rotate plane polarised light.

  (a) Draw the structural formula of the structural isomer of C_4H_9OH which contains an asymmetric centre.

  [1]

  (b) Draw the 3D structures of two optical isomers of the molecule identified in (a).

  [2]

  (c) Explain the term **plane polarised light.**

  _____

  _____ [1]

  (d) How can a solution of one optical isomer be distinguished from a solution of the other optical isomer of a substance?

  _____

  _____ [2]

  (e) Only one isomer of C_4H_9OH can be oxidised to a ketone using acidified potassium dichromate(VI) solution. Draw the alcohol and the corresponding ketone.

  [2]

4.7 Aldehydes and Ketones

Naming and Physical Properties

1 Give the IUPAC name for each of the organic compounds below.

(a)	butanal (H-C-C-C-CHO structure)	(b)	propanone
(c)	hexan-2-one like structure (C=O on C2 of 6-carbon chain)	(d)	2-methylbutanal
(e)	CH₃CHO	(f)	CH₃CH₂CHClCOCH₃
(g)	1-bromo-4-methylpentan-2-one	(h)	4-hydroxybutanal
(i)	pentane-2,4-dione	(j)	2,4-dichloro-3-methylhexanal-type

[10]

2 Explain why propanal and propanone are both soluble in water.

_____

_____ [2]

3 Suggest why the boiling point of 2-methylpropanal is lower than that of its structural isomer butanal.

_____

_____ [2]

4 Ethanal has a boiling point of 21 °C which is very close to room temperature. When ethanal is prepared it is separated from the reaction mixture by distillation and collected. Suggest how the ethanal could be collected.

_____

_____ [1]

5 Explain the difference in boiling points between the three substances in the table.

Substance	Relative formula mass (Mr)	Boiling point / °C
$CH_3CH_2CH_3$ (propane)	44	−42
CH_3CHO (ethanal)	44	+21
CH_3CH_2OH (ethanol)	46	+79

_____

_____

_____

_____ [3]

6 Draw a diagram to show the hydrogen bonding which occurs between a molecule of ethanal and water. Include lone pairs and partial charges.

[2]

Oxidation and reduction of aldehydes and ketones

7 Cinnamaldehyde is a food additive used in chewing gum and cake mixes to give a cinnamon flavour. Its skeletal formula is given below.

(a) State and explain which geometric isomer of cinnamaldehyde is shown.

_____

_____ [1]

(b) Describe how you could use two different chemical tests to prove that cinnamaldehyde is an aldehyde.

_____

_____

_____

_____

_____ [4]

(c) Write an equation for the oxidation of cinnamaldehyde using [O] as the oxidant. Name an oxidising agent which could be used.

_____

_____ [2]

8 Which one of the following compounds forms optically active compounds on reduction using lithal?

   A $CH_3CH_2C(CH_3)=CHCH_3$

   B $CH_3CH_2C(CH_3)=CH_2$

   C CH_3COCH_3

   D $CH_3CH_2COCH_3$ [1]

9 Which one of the following alcohols could **not** be produced by the reduction of an aldehyde or a ketone?

    A   2-methylpentan-1-ol

    B   2-methylpentan-2-ol

    C   3-methylpentan-1-ol

    D   3-methylpentan-2-ol   [1]

10 Draw the structural formula and give the IUPAC name for the organic products (if any) of the reactions below.

(a) CH_3CHO with $LiAlH_4$

_____ [2]

(b) CH_3COCH_3 with $LiAlH_4$

_____ [2]

(c) CH_3CHO with ammoniacal silver nitrate

_____ [2]

(d) $CH_3CH(OH)CH_3$ with acidified potassium dichromate(VI) solution

_____ [2]

4.7 ALDEHYDES AND KETONES

11 Several compounds, such as A and B, have the molecular formula $C_5H_{10}O_2$.

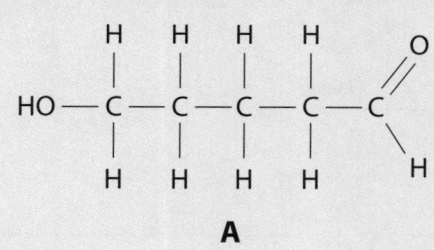

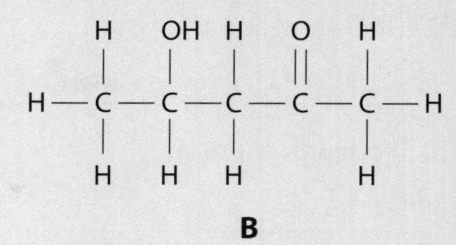

A: HO—CH₂—CH₂—CH₂—CH₂—CHO

B: CH₃—CH(OH)—CH₂—CO—CH₃

(a) Name the functional groups present in A and in B.

A _____

B _____ [2]

(b) Draw the structural formula of the organic products, if any, which would be formed on treatment of A and B with excess of the reagent given below. State 'no reaction', if you predict no reaction would occur.

(i) With Fehling's solution

A   B

[2]

(ii) With acidified potassium dichromate(VI) solution

A   B

[2]

(c) Write the ionic equation for the reduction of the metal ion in each of the following.

   (i) Tollens' reagent

   _____ [1]

   (ii) Fehling's solution

   _____ [1]

12 Compound C can be converted into compound D.

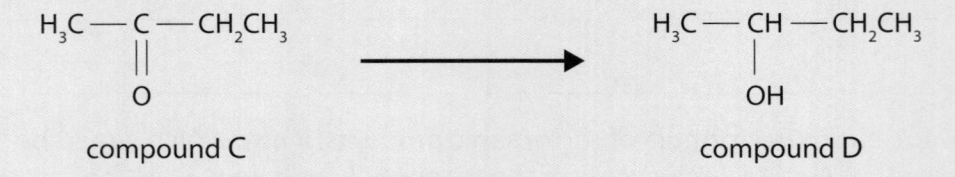

$$H_3C-\underset{\underset{O}{\|}}{C}-CH_2CH_3 \longrightarrow H_3C-\underset{\underset{OH}{|}}{CH}-CH_2CH_3$$

   compound C   compound D

   (a) Give the IUPAC names for each of the following.

   Compound C _____

   Compound D _____ [2]

   (b) Name a reagent used to convert C to D.

   Reagent: _____ [1]

   (c) Name the organic product when D is warmed with acidified potassium dichromate(VI) solution.

   _____ [1]

13 Aldehydes can be distinguished from ketones using the reagents shown in the table.

   (a) Complete the following table.

Reagent	Formula of metal/ion before test	Formula of metal/ion after test
Tollens' reagent		
Fehling's solution		

[2]

4.7 ALDEHYDES AND KETONES

(b) Describe, giving practical details, how you would carry out a test tube reaction using Fehling's solution to prove that a liquid is propanone and not propanal. State the observations which occur.

_____

_____

_____

_____

_____ [4]

Nucleophilic addition reactions

14 Butanone reacts with hydrogen cyanide.

(a) Write the equation for the reaction of hydrogen cyanide with butanone and name the organic product.

_____ [2]

(b) Name and outline the mechanism for the reaction of butanone with hydrogen cyanide.

[4]

(c) By considering the mechanism of the reaction explain why the product of this reaction is optically inactive.

_____

_____

_____

_____

_____ [4]

15 The taste of apples is due to the compound $CH_3CH_2CH_2CH=CHCHO$.

(a) Draw the structural formula of the two geometric isomers of this compound and give their IUPAC names.

[3]

(b) Write an equation for the reaction of this compound with bromine water.

_____ [1]

(c) Draw the structure of the compound formed when $CH_3CH_2CH=CHCHO$ reacts with hydrogen cyanide.

[1]

(d) Name a reagent which can be used to oxidise CH₃CH₂CH₂CH=CHCHO and give the structure of the product.

_____ [2]

16 Compound E (CH₃COCH₂CH₂CH₃) and compound F (CH₃CH₂COCH₂CH₃) are isomers.

(a) Explain why E and F are isomers

_____ [1]

(b) Give the IUPAC name of E.

_____ [1]

(c) Name the mechanism for the reaction of compound E with HCN and name the product formed.

_____ [2]

(d) Explain why a test of the optical activity of the products formed when E and F react with HCN would not distinguish between them.

_____ [3]

(e) Describe what is observed if E and F are heated in separate test tubes with Fehling's solution.

_____ [2]

17 Aldehydes contain the carbonyl functional group.

   (a) On the diagram below show the polarity of the carbonyl group.

   C═O

   [1]

   (b) Nucleophiles react with carbonyl compounds to form addition products. Define the term **nucleophile**.

   _____
   _____ [2]

   (c) Explain why the carbonyl group is susceptible to attack by a nucleophile.

   _____
   _____ [1]

   (d) Ethanal reacts with hydrogen cyanide. Write an equation for this reaction and name the organic product.

   _____
   _____ [2]

   (e) Name and outline the mechanism for the reaction of ethanal with hydrogen cyanide.

   _____ [4]

4.7 ALDEHYDES AND KETONES

Reaction with 2,4-dinitrophenylhydrazine

18 In diabetic people propanone is produced during the breakdown of fats in the body. Propanone reacts with 2,4-dinitrophenylhydrazine.

(a) Write the equation for the reaction of propanone with 2,4-dinitrophenylhydrazine.

[3]

(b) Name the product of the reaction in (a) and state its appearance.

[2]

(c) Explain how a sample of propanone liquid could be identified using 2,4-dinitrophenylhydrazine. Include experimental details in your answer. You do not need to give details of recrystallisation.

In this question you will be assessed on using your written communication skills including the use of specialist scientific terms.

[6]

(d) Recrystallisation is a method of purifying solids. Describe how recrystallistion is carried out and describe factors which need to be considered when choosing a solvent for recrystallisation.

_____

_____

_____

_____

_____

_____ [4]

(e) Calculate the volume of propanone needed to produce 10 g of the product in reaction (a) assuming an 80 % yield. The density of propanone is 0.790 g cm^{-3}. Give your answer to one significant figure.

_____

_____

_____

_____

_____

_____ [4]

(f) Describe, giving practical details how a melting point determination can be carried out to identify the product from the reaction in (a).

_____

_____

_____

_____

_____ [4]

4.7 ALDEHYDES AND KETONES

19 Write equations for the reaction of each of the following compounds with 2,4-dinitrophenylhydrazine.

   (a) butanone

[3]

   (b) benzaldehyde (structure given below)

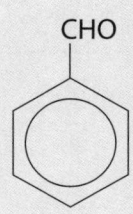

[3]

4.8 Carboxylic Acids

Nomenclature and physical properties

1. Give the IUPAC name for each of the following.

   (a) CH₃CH₂CClHCOOH _____ [1]

   (b) HCOOH _____ [1]

   (c) (ethanoic acid structure) _____ [1]

   (d) (propanoic acid structure) _____ [1]

   (e) (butanedioic acid structure) _____ [1]

   (f) (3-hydroxypentanoic acid structure) _____ [1]

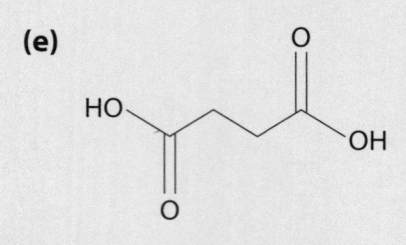

2. Write the molecular formula for each compound in question 1.

   (a) _____ [1]

   (b) _____ [1]

   (c) _____ [1]

   (d) _____ [1]

   (e) _____ [1]

   (f) _____ [1]

4.8 CARBOXYLIC ACIDS

3 (a) Draw the structure of the following carboxylic acids.

    (i) 3-chloropentanoic acid

[1]

    (ii) 4-methylpent-2-enoic acid

[1]

    (iii) 2-hydroxypropanoic acid

[1]

    (iv) 3-hydroxypropanoic acid

[1]

(b) Three of the acids in (a) can exist as stereoisomers. Which one cannot?

_____ [1]

(c) Name the acid in (a) which can exist as E-Z isomers.

_____ [1]

(d) Explain why propanoic acid is soluble in water.

_____

_____ [2]

4 Draw a diagram to show two hydrogen bonds between one ethanoic acid molecule and two water molecules.

[2]

5 Suggest why the boiling point of ethanoic acid is higher than propan-1-ol

Substance	Formula	Relative formula mass (Mr)	Boiling point / °C
Propan-1-ol	$CH_3CH_2CH_2OH$	60	97.2
Ethanoic acid	CH_3COOH	60	118.0

_____
_____
_____
_____ [2]

Preparation of carboxylic acids

6 Which one of the following reactions will **not** form propanoic acid?

   A   Acid catalysed hydrolysis of ethyl propanoate

   B   Acid catalysed hydrolysis of propyl ethanoate

   C   Acid catalysed hydrolysis of propanenitrile

   D   Oxidation of propan-1-ol [1]

7 Write equations for the preparation of propanoic acid by the following reactions.

   (a) oxidation of propanal

   _____ [1]

(b) oxidation of propan-1-ol

_____ [1]

(c) hydrolysis of ethyl propanoate using acid

_____ [1]

(d) hydrolysis of propanenitrile using hydrochloric acid

_____ [1]

(e) hydrolysis of propanenitrile using sodium hydroxide solution followed by hydrochloric acid

_____ [2]

8 Butanoic acid can be prepared from butan-1-ol and potassium dichromate (VI).

(a) Write an equation for this reaction.

_____ [1]

(b) Describe how you would experimentally prepare a sample of butanoic acid.
In this question you will be assessed on using your written communication skills including the use of specialist scientific terms.

_____

_____

_____

_____

_____

_____

_____

_____

_____ [6]

Reactions of acids

9 (a) Write an equation for the dissociation of ethanoic acid in solution.

_____ [1]

(b) Name the anion formed in this reaction.

_____ [1]

(c) Would you expect a solution of ethanoic acid to conduct electricity? Explain your answer.

_____ [1]

10 Write equations and give observations for the following reactions of carboxylic acids.

(a) propanoic acid and ammonia

_____

_____ [2]

(b) ethanoic acid and sodium carbonate

_____

_____ [2]

(c) butanoic acid and sodium hydroxide

_____

_____ [2]

11 Name all the products in the reactions in question 10.

(a) _____ [1]

(b) _____ [1]

(c) _____ [1]

12 Write equations for the formation of the following salts.

(a) potassium ethanoate from potassium hydroxide

_____ [1]

(b) sodium ethanoate from sodium carbonate

_____ [1]

(c) magnesium propanoate from magnesium

_____ [1]

(d) ammonium methanoate from ammonia

_____ [1]

13 Write equations for the following reactions

(a) ethanoic acid + PCl$_5$

_____ [1]

(b) butanedioic acid + excess PCl$_5$

_____ [1]

(c) reduction of propanoic acid using lithal ([H])

_____ [1]

14 Name the products in the reaction in question 13(a) and state the observations made.

_____

_____

_____ [2]

15 Some reactions of propanoic acid are shown below. Draw structure of the products of each reaction.

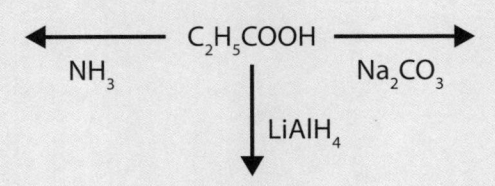

$$\xleftarrow{NH_3} C_2H_5COOH \xrightarrow{Na_2CO_3}$$
$$\downarrow LiAlH_4$$

[3]

16 Salicylic acid is used to make aspirin. Its structure is shown below.

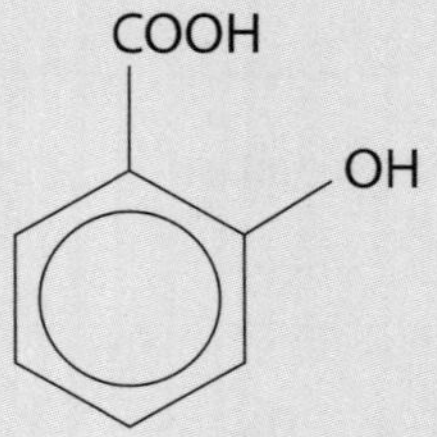

[Structure: benzene ring with COOH and OH (ortho)]

(a) Write the equation for the reaction of salicylic acid with phosphorus(V) chloride. The OH group bonded directly to the benzene ring does not react.

[1]

(b) Write an equation for the reaction of salicylic acid with lithal.
(Use [H] to represent lithal.)

[1]

17 The scheme below summarises the formation and reactions of ethanoic acid.

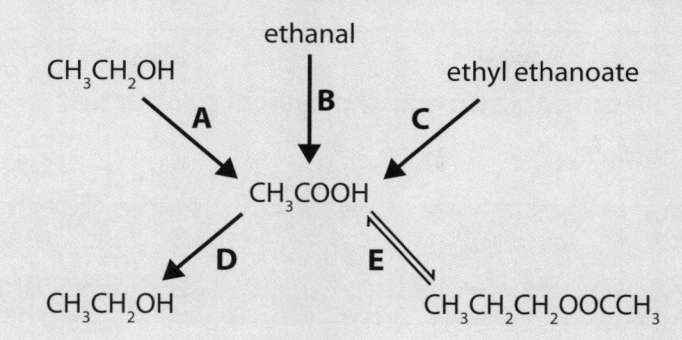

$$CH_3CH_2OH \xrightarrow{A} CH_3COOH \xleftarrow{B}\text{ethanal} \xleftarrow{C} \text{ethyl ethanoate}$$

$$CH_3COOH \xrightarrow{D} CH_3CH_2OH \qquad CH_3COOH \rightleftharpoons_{E} CH_3CH_2CH_2OOCCH_3$$

(a) Give suitable reagents and conditions for each of the reactions A–E.

A _____ [1]

B _____ [1]

C _____ [1]

D _____ [1]

E _____ [1]

(b) Give the type of reaction occurring in the reactions A–E.

A _____ [1]

B _____ [1]

C _____ [1]

D _____ [1]

E _____ [1]

4.9 Derivatives of Carboxylic Acids

Nomenclature and physical properties

1. Complete the table below.

Carboxylic acid	Alcohol	Ester name	Ester structural formula	Ester molecular formula
ethanoic acid	ethanol			
ethanoic acid	methanol			
propanoic acid	ethanol			
butanoic acid	propan-1-ol			

[4]

2. Give the IUPAC name for the following esters.

   (a) 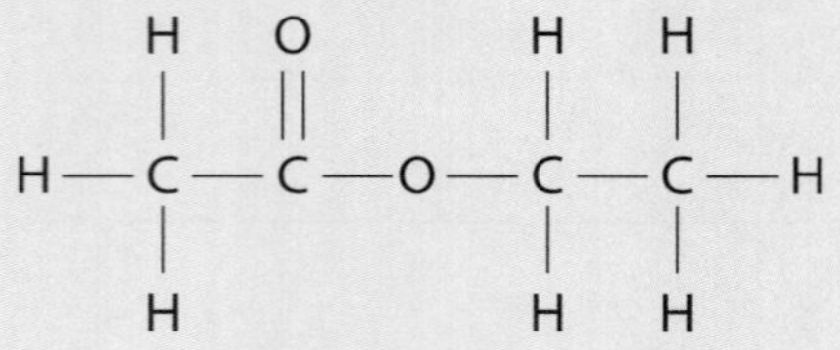

   H—C(H)(H)—C(=O)—O—C(H)(H)—C(H)(H)—H

   _____ [1]

   (b) 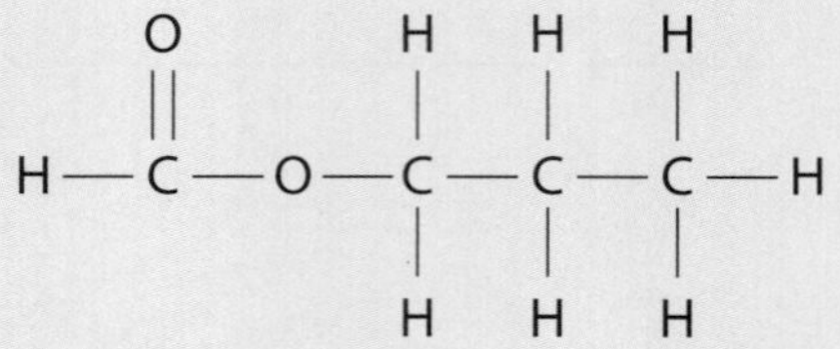

   H—C(=O)—O—C(H)(H)—C(H)(H)—C(H)(H)—H

   _____ [1]

   (c) $CH_3CH_2OOCH_2CH_3$ _____ [1]

   (d) $HCOOCH_3$ _____ [1]

   (e) $CH_3CH_2OOCCH_2CH_3$ _____ [1]

   (f) $HCOOCH_2CH_3$ _____ [1]

4.9 DERIVATIVES OF CARBOXYLIC ACIDS

3 Some esters and their solubility in water are shown in the table below.

Ester	Formula	Solubility (g per 100 g of water)
ethyl methanoate	HCOOCH$_2$CH$_3$	10.5
ethyl ethanoate	CH$_3$COOCH$_2$CH$_3$	8.7
ethyl propanoate	CH$_3$CH$_2$COOCH$_2$CH$_3$	1.7

(a) Explain why ethyl methanoate is soluble in water.

_____

_____ [2]

(b) State and explain the trend in solubility shown in the table.

_____

_____ [2]

(c) Explain why the boiling point of ethyl methanoate (53 °C) is lower than the boiling point of ethyl propanoate (102 °C).

_____

_____

_____ [2]

4 Propanoic acid is used as a preservative and flavouring agent in packaged foods such as baked goods and cheese. Calcium propanoate is a salt of propanoic acid which is used in bread to prevent mould.

(a) Write an equation for the reaction of propanoic acid and calcium carbonate.

_____ [1]

(b) (i) Draw and name two structural isomers of propanoic acid which are esters.

[2]

95

(ii) Explain why the boiling points of the ester isomers of propanoic acid are lower than the boiling point of propanoic acid.

_____

_____

_____

_____ [4]

5 $CH_3C(CH_3)HCH_2COOC_2H_5$ is an example of what type of organic compound.

    A   aldehyde

    B   carboxylic acid

    C   ester

    D   ketone   [1]

Acyl chlorides

6 Give the IUPAC name for the following:

    (a) CH_3COCl _____ [1]

    (b) $HCOCl$ _____ [1]

    (c) $CH_3CH_2CH_2COCl$ _____ [1]

7 Write equations and give observations for the following reactions.

    (a) ethanoyl chloride + water

_____

_____ [2]

    (b) propanoyl chloride + ethanol

_____

_____ [2]

    (c) Describe a chemical test for the inorganic product of the reaction, in (a) and (b) indicating a positive result for this test.

_____

_____ [2]

(d) Write an equation for the reaction of excess ethanoyl chloride with propane-1,3-diol.

_____ [1]

Preparation of esters

8 Which one of the following statements about the formation of an ester from ethanoyl chloride and propan-1-ol is correct?

 A Concentrated sulfuric acid is required.

 B Heat is required.

 C The ester produced is called ethyl propanoate.

 D The reaction goes to completion [1]

9 Write equations for the formation of each of the following esters from an alcohol and a carboxylic acid.

(a) methyl ethanoate

_____ [1]

(b) ethyl pentanoate

_____ [1]

(c) propyl butanoate

_____ [1]

10 Simple monoesters, such as the one shown below, can be prepared by reacting carboxylic acids with alcohols.

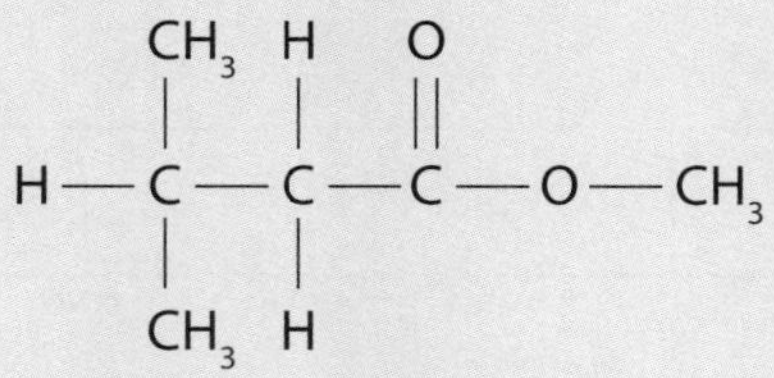

$$\begin{array}{c} \quad\quad CH_3\ \ H\ \ \ O \\ \quad\quad\ \ |\ \ \ \ \ |\ \ \ \ \ || \\ H-C-C-C-O-CH_3 \\ \quad\quad\ \ |\ \ \ \ \ | \\ \quad\quad CH_3\ \ H \end{array}$$

(a) Give the IUPAC names of the carboxylic acid and alcohol required to prepare this ester.

_____

_____ [2]

(b) Write the equation for the formation of the ester.

_____ [1]

(c) The ester is formed by heating the reactants with concentrated sulfuric acid under reflux.

   (i) Explain why anti-bumping granules are added to the reflux flask.

   _____ [1]

   (ii) State two roles of the concentrated sulfuric acid in this reaction.

   _____
   _____ [2]

   (iii) Describe what happens when a reaction mixture is refluxed.

   _____
   _____ [1]

   (iv) Suggest and explain how the reflux flask is heated.

   _____
   _____ [2]

(d) After reflux the ester is separated from the reaction mixture and purified.

   (i) Describe how a sample of the ester is obtained. Include a description of the apparatus used.

   _____
   _____
   _____ [3]

   (ii) Describe how the ester is purified using sodium carbonate solution.

   _____
   _____
   _____ [3]

(iii) Describe how the ester is dried.

_____

_____

_____ [3]

(iv) State how the ester could be tested to determine its purity.

_____

_____ [2]

(v) State a theoretical reason why the reaction yield of this reaction is less than 100%.

_____

_____ [1]

11 Butan-1-ol was reacted with an excess of propanoic acid in the presence of a small amount of concentrated sulfuric acid. 6.0 g of the alcohol produced 7.4 g of the ester. Which one of the following is the percentage yield of the ester?

  A   57%

  B   70%

  C   75%

  D   81% [1]

12 Ethyl butanoate is an ester which has the flavour similar to that of orange juice and is commonly used as an artificial flavouring.

(a) Write the equation for the preparation of ethyl butanoate from an acid and an alcohol and name the catalyst used in the reaction.

_____

_____ [2]

(b) Write the equation for the preparation of ethyl butanoate from an acyl chloride and an alcohol.

_____ [1]

(c) State three advantages of using an acyl chloride instead of a carboxylic acid in esterification reactions.

_____

_____

_____ [3]

13 The flow scheme below shows a series of organic reactions.

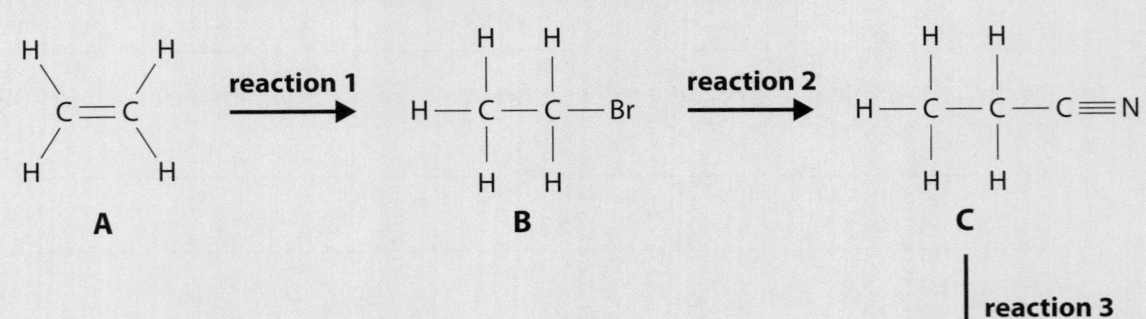

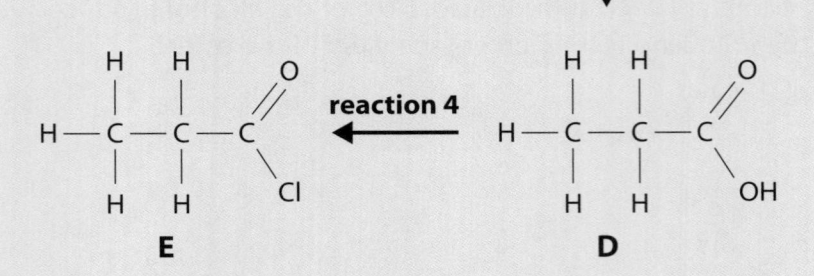

(a) Name compounds A to E.

A _____

B _____

C _____

D _____

E _____ [5]

(b) State the type of reaction which is taking place in reactions 1 and 2.

Reaction 1: _____

Reaction 2: _____ [2]

4.9 DERIVATIVES OF CARBOXYLIC ACIDS

(c) Name the mechanism by which reaction 2 occurs.

_____ [1]

(d) Name the reagents required for each of reactions 2 and 3.

Reaction 2: _____

Reaction 3: _____ [2]

(e) Name a reagent which could be used for reaction 4.

_____ [1]

14 For each of the following pairs of compounds, name a suitable reagent or reagents that could be added separately to a test tube of each compound in order to distinguish between them. Describe what you would observe with each compound.

(a) Methyl ethanoate and propanoic acid.

_____

_____

_____ [3]

(b) Butan-2-ol and 2-methylpropan-2-ol

_____

_____

_____ [3]

Fats and oils

15 (a) Draw the structure and give the IUPAC name for glycerol.

_____ [1]

(b) Write an equation for the reaction of glycerol with three molecules of the stearic acid CH$_3$(CH$_2$)$_{16}$COOH to form a fat.

[2]

(c) Write an equation for the hydrolysis of Fat A, shown below, with excess sodium hydroxide.

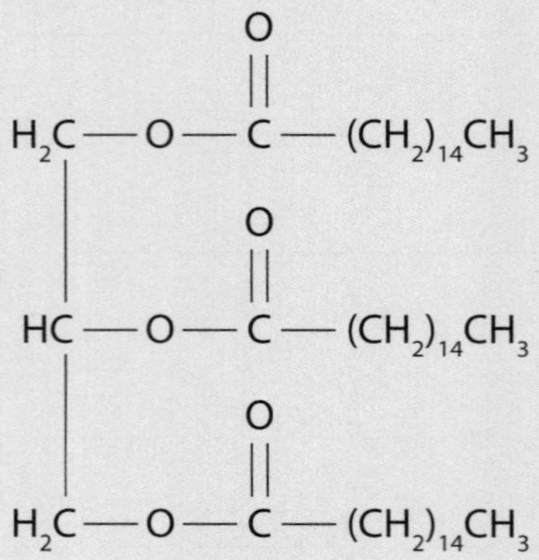

H$_2$C—O—C(=O)—(CH$_2$)$_{14}$CH$_3$
|
HC—O—C(=O)—(CH$_2$)$_{14}$CH$_3$
|
H$_2$C—O—C(=O)—(CH$_2$)$_{14}$CH$_3$

[1]

(d) Fat A can be used to make biodiesel. Suggest the conditions used and write an equation.

Conditions: _____ [4]

(e) Define the term biodiesel.

_____

_____ [1]

(f) Define the term transesterification.

_____

_____

_____ [2]

16 Hydrolysis of an oil produces 1 mole of propane-1,2,3-triol and 3 moles of the sodium salt of octadeca-9,12-dienoic acid as the only products.

$$CH_3(CH_2)_4CH=CHCH_2CH=CH(CH_2)_7COOH$$
octadeca-9,12-dienoic acid

(a) Draw the structural formula of propan-1,2,3-triol and give its common name.

_____ [2]

(b) Draw the structure of the oil.

[1]

(c) State the meaning of the term hydrolysis and state the conditions for this hydrolysis.

_____

_____

_____ [2]

17 The fat shown below, stearin, is present in lard.

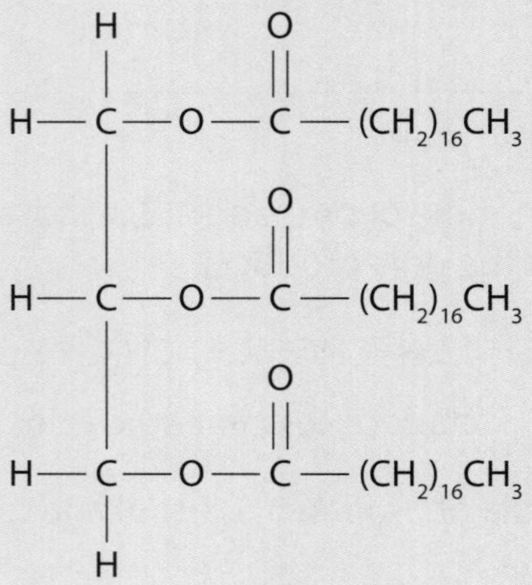

$$\begin{array}{c} H \\ | \\ H-C-O-C(=O)-(CH_2)_{16}CH_3 \\ | \\ H-C-O-C(=O)-(CH_2)_{16}CH_3 \\ | \\ H-C-O-C(=O)-(CH_2)_{16}CH_3 \\ | \\ H \end{array}$$

(a) When refluxed with potassium hydroxide, stearin produces the potassium salt of stearic acid (potassium stearate) and only one other product.

(i) Write an equation for the reaction of stearin with excess potassium hydroxide.

[2]

(ii) State the IUPAC name of the other product.

_____ [1]

(b) The methyl ester of stearic acid is present in biodiesel.

(i) Give the molecular formula of the methyl ester.

_____ [1]

(ii) Write the equation for the complete combustion of this ester.

_____ [1]

4.9 DERIVATIVES OF CARBOXYLIC ACIDS

18 Ester X, shown below, can be used to make biodiesel using transesterification. This ester is formed when glycerol reacts with palmitic acid.

$$\begin{array}{l} CH_2OCOC_{15}H_{31} \\ | \\ CHOCOC_{15}H_{31} \\ | \\ CH_2OCOC_{15}H_{31} \end{array}$$

Ester X

(a) Write an equation for the formation of X from glycerol and palmitic acid.

[1]

(b) Explain why X is a saturated fat.

_____

_____ [2]

(c) Write the equation for the reaction between X and methanol.

[1]

(d) Write the formula of the product from reaction (c) that could be used as biodiesel.

_____ [1]

105

19 One of the fats in rape seed oil is trilinolein and it may be represented as:

CH₂OOC(CH₂)₇CH=CHCH₂CH=CH(CH₂)₄CH₃
|
CHOOC(CH₂)₇CH=CHCH₂CH=CH(CH₂)₄CH₃
|
CH₂OOC(CH₂)₇CH=CHCH₂CH=CH(CH₂)₄CH₃

(a) Write an equation for the formation of trilinolein from 3 molecules of the fatty acid, linoleic acid, and glycerol.

[2]

(b) Explain why trilinolein is described as an unsaturated fat.

_____

_____ [1]

(c) Trilinolein can be converted into biodiesel by reaction with methanol. Draw the structure of this biodiesel molecule.

[1]

(d) The vegetable oil shown below reacts with methanol in the presence of potassium hydroxide to form biodiesel. Write the equation for this reaction.

CH₂OOCC₁₇H₃₁
|
CHOOOC₁₇H₃₃
|
CH₂OOCC₁₇H₂₉

[2]

4.9 DERIVATIVES OF CARBOXYLIC ACIDS

20 Dodecanoic acid, $C_{11}H_{23}COOH$, is a white solid at room temperature with a melting point of 45 °C. It is insoluble in water.

(a) Explain why ethanoic acid is soluble in water whereas dodecanoic acid is insoluble.

_____

_____ [2]

(b) State what is observed when sodium carbonate solution reacts with dodecanoic acid.

_____

_____ [1]

(c) Describe how you would experimentally determine that a sample of solid dodecanoic acid is pure.

_____

_____

_____

_____ [4]

(d) Dodecanoic acid can be reduced to the corresponding alcohol. Write an equation for the reduction using [H] to represent the reducing agent and name a suitable reducing agent for the reaction.

_____ [2]

(e) A triester can be formed from dodecanoic acid and glycerol. Write an equation for the formation of this triester.

[1]

4.10 Aromatic Chemistry

Structure of benzene

1 (a) Give the molecular formula and the empirical formula of benzene.

_____ [1]

(b) State the shape of benzene and give the bond angle around a carbon atom.

_____
_____ [2]

(c) State the number of delocalised electrons in a molecule of benzene.

_____ [1]

(d) State the IUPAC name for the following.

(i) methylbenzene

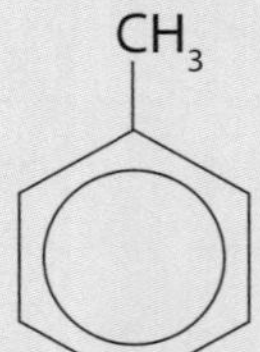

(ii) nitrobenzene

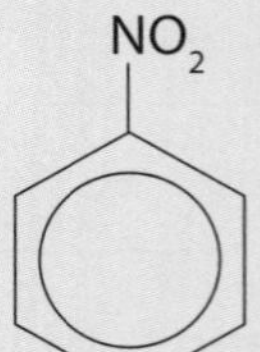

(iii) 1,2-dichlorobenzene

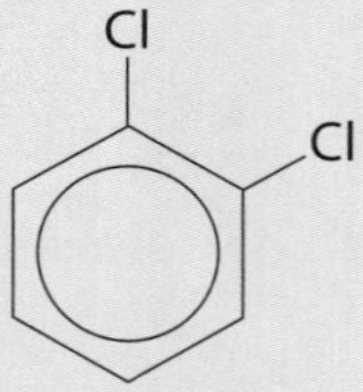

(iv) 1-ethyl-2-methylbenzene

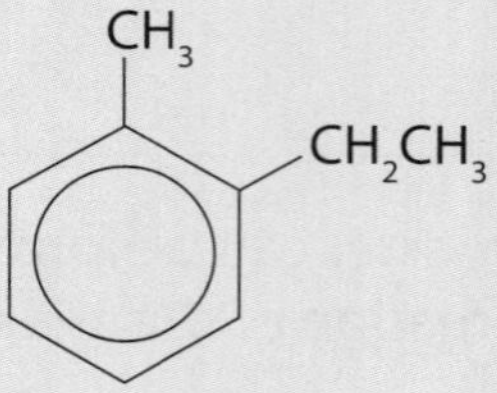

(v) 1-phenylpropan-1-one

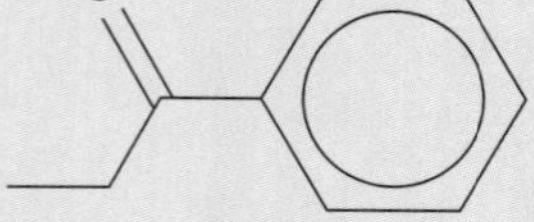

(vi) 1-bromo-4-chlorobenzene

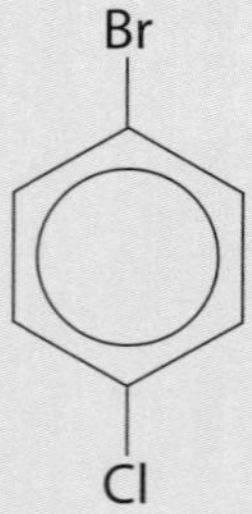

[6]

(e) How does a delocalised ring of electrons form in benzene?

_____

_____ [2]

(f) Draw a possible structure for an aromatic compound which has the empirical formula C₇H₈.

[1]

(g) How many sigma bonds are present in a benzene molecule?

_____ [1]

(h) What is meant by delocalisation in terms of benzene?

_____ [1]

2 (a) Name a reagent which could be used to distinguish between benzene and cyclohexene.

_____ [1]

(b) State what is observed when this reagent is shaken with benzene and then with cyclohexene in separate test tubes.

_____

_____ [1]

3 How many aromatic isomers of dichlorobenzene exist?

A 1
B 2
C 3
D 4 [1]

4 Explain the bonding in and the shape of a benzene molecule. Explain why it undergoes substitution reactions with bromine more readily than addition reactions.

In this question you will be assessed on using your written communication skills including the use of specialist scientific terms.

_____
_____
_____
_____
_____
_____
_____
_____
_____ [6]

5 Which one of the following does not contain any delocalised electrons?

A   poly(chlorethene)

B   benzene

C   graphite

D   zinc [1]

6 Which one of the following is the total number of electrons involved in bonding in benzene?

A   12

B   18

C   24

D   30 [1]

4.10 AROMATIC CHEMISTRY

7 Which one of the following is the relative molecular mass of 1,2-dichlorobenzene?

   A   147

   B   149

   C   151

   D   153 [1]

8 Compare the length of the bond between carbon atoms in benzene with that in ethene.

_____

_____

_____ [1]

Electrophilic substitution

9 What is the name of the NO_2^+ ion?

   A   nitrate cation

   B   nitrite cation

   C   nitronium ion

   D   nitrosonium ion [1]

10 2-nitrotoluene is a yellow liquid which is used in many dyes. It is produced by nitration of toluene.

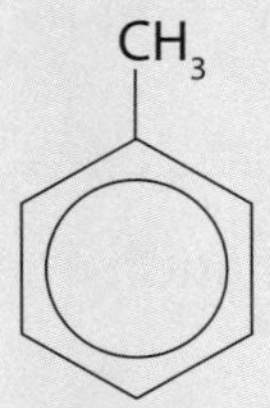

Toluene

(a) Name the type of mechanism for the production of 2-nitrotoluene.

_____ [1]

(b) Outline the mechanism for the reaction in (a).

[5]

11 Methyl 3-nitrobenzoate exists as a solid at room temperature. Its melting point is 78–79 °C.

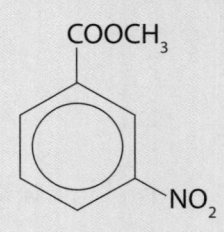

(a) State the functional group in methyl 3- nitrobenzoate which can be hydrolysed.

_____ [1]

(b) (i) Write the equation for the formation of methyl 3-nitrobenzoate from methyl benzoate using nitric acid.

[1]

(ii) Assuming a 60% yield, calculate the minimum mass of methyl benzoate required to produce 5.43 g of methyl 3-nitrobenzoate when the other reagents are in excess.

_____

_____

_____

_____ [3]

(iii) Name the two acids in the nitrating mixture used in the formation of methyl 2-nitrobenzoate. Write an equation to show how these two acids react when mixed.

_____

_____ [2]

(iv) Name the ion, produced in the reaction in (c)(i) which attacks the methyl benzoate molecule.

_____ [1]

(v) Name and outline the mechanism of the mononitration of methyl benzoate and name the mechanism.

Name: _____ [4]

(d) In the preparation the crude methyl 2-nitrobenzoate is recrystallised by dissolving in the minimum volume of hot solvent. The product is filtered off using filtration under reduced pressure and washed using a small amount of cold solvent before drying.

(i) Explain why recrystallisation is carried out and why a minimum volume of hot solvent was used.

_____

_____

_____ [2]

(ii) Name three pieces of equipment which are essential for filtration under reduced pressure.

_____

_____ [2]

(iii) State why the product was washed with cold solvent.

_____ [1]

(iv) Describe how the product could be dried.

_____ [1]

(v) State how the melting point is used to determine whether the crystals are pure methyl 3-nitrobenzoate.

_____

_____ [2]

(vi) Explain why drying is necessary before carrying out a melting point determination.

_____ [1]

12 2-methylpropanoyl chloride can be used to acylate benzene in part of a series of reactions used to produce ibuprofen.

$$H_3C - \underset{\underset{CH_3}{|}}{\overset{\overset{H}{|}}{C}} - \overset{\overset{O}{\|}}{C} - Cl$$

2-methylpropanoyl chloride

(a) Draw the skeletal formula of 2-methylpropanoyl chloride.

[1]

(b) (i) The reaction of 2-methylpropanoyl chloride with benzene is a substitution reaction. Explain why it is described as a substitution reaction.

_____

_____ [1]

4.10 AROMATIC CHEMISTRY

(iii) Identify the catalyst required in this reaction. Write equations to show how the catalyst is used to form a reactive intermediate and how the catalyst is reformed at the end of the reaction.

_____

_____

_____

_____ [3]

(iv) Name and outline a mechanism for the reaction of benzene with this reactive intermediate.

_____ [4]

15 In which of the following is the halogen acting as an electrophile?

A $CH_3CH_3 + Cl_2 \rightarrow CH_3CH_2Cl + HCl$

B $CH_4 + 4Cl_2 \rightarrow CCl_4 + 4HCl$

C $C_6H_6 + Br_2 \rightarrow C_6H_5Br + HBr$

D $C_6H_5CH_3 + Br_2 \rightarrow C_6H_5CH_2Br + HBr$ [1]

16 The flow scheme below shows some organic conversions.

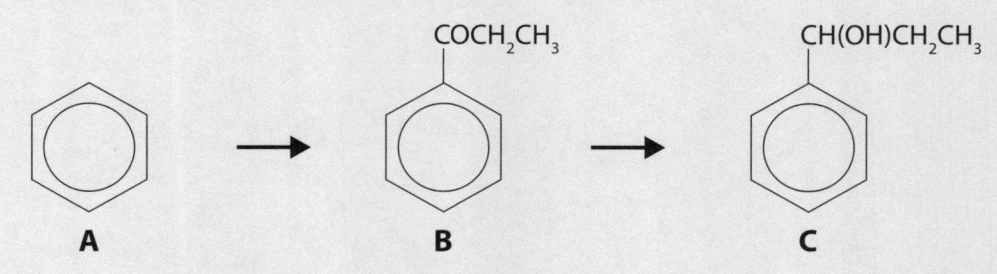

A (benzene) → B (C$_6$H$_5$COCH$_2$CH$_3$) → C (C$_6$H$_5$CH(OH)CH$_2$CH$_3$)

(a) Name compound B

_____ [1]

115

(b) Name the mechanism for the reaction when A is converted into B and suggest a suitable reagent or combination of reagents for the reaction.

_____

_____ [2]

(c) The conversion of B to C is a reduction reaction. Suggest a suitable reagent or combination of reagents for the reaction. State if the alcohol formed is primary or secondary.

_____

_____ [2]

(d) Explain which compound A, B, or C will form optical isomers.

_____

_____ [1]

17 Which one of the following describes the appearance of methyl 3-nitrobenzoate?

- A   Colourless liquid
- B   Cream solid
- C   Orange solid
- D   Violet crystals [1]

Unit A2 2:
Analytical, Transition Metals, Electrochemistry and Organic Nitrogen Chemistry

5.1 Mass Spectrometry

1 A sample of gas was analysed using a mass spectrometer. The molecular ion was detected at a mass to charge ratio of 28 in a low resolution mass spectrometer and 28.0312 in a high resolution mass spectrometer.

   (a) Which two of the following gases could be present in the sample?

   ethane   ethene   carbon dioxide   carbon monoxide

   _____ [1]

   (b) Use the precise relative atomic masses below to explain how high resolution mass spectrometry can be used to identify the gas in the sample.

   ^{1}H = 1.0078 ^{16}O = 15.9949 ^{14}N = 14.0071 ^{12}C = 12.0000

   _____

   _____

   _____

   _____ [3]

2 A molecular ion peak of a compound was found at a mass to charge ratio of 150.1041. Which of the following is the molecular formula of the compound? Use the precise relative atomic masses given in question 1(b).

   A $C_9H_{10}O_2$

   B $C_{11}H_{18}$

   C $C_{10}H_{14}$

   D $C_{10}H_{14}O$ [1]

3 A simple mass spectrum of ethanol is shown below.

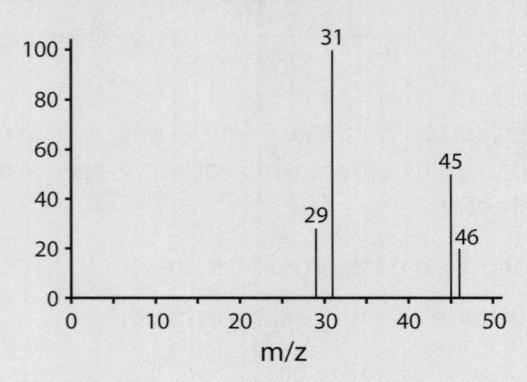

(a) What label should be placed on the *y*-axis?

_____ [1]

(b) What is meant by the term **m/z**?

_____ [1]

(c) State the m/z ratio of the molecular ion.

_____ [1]

(d) Define the term **molecular ion peak**.

_____

_____ [2]

(e) Define the term base peak and identify the m/z value of the base peak in this spectrum.

_____

_____

_____ [3]

4 Chlorine has two isotopes, chlorine-35 and chlorine-37.

Which one of the following is the number of peaks found in the mass spectrum of chlorine gas?

A 2

B 3

C 4

D 5

[1]

5.1 MASS SPECTROMETRY

5 The mass spectrum of a compound is shown below. Which one of the following is the base peak?

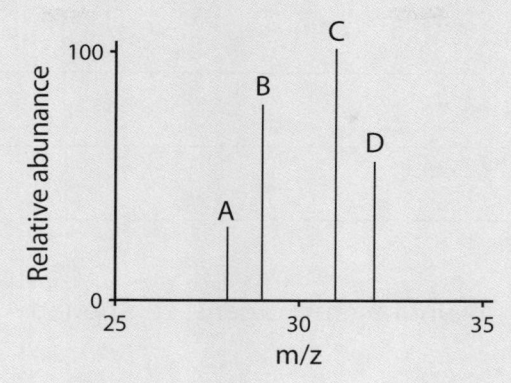

[1]

6 The mass spectrum of the ester methyl propanoate, $CH_3CH_2COOCH_3$, is shown below.

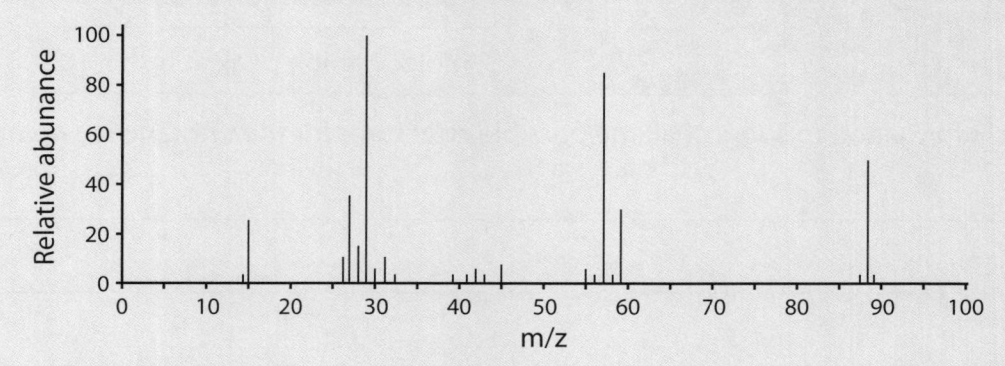

(a) State the m/z value of the base peak.

_____ [1]

(b) Explain why there is a peak at 89.

_____
_____ [2]

(c) The mass spectrum shows peaks due to fragmentation ions.

  (i) Define the term **fragmentation ion**.

  _____
  _____ [1]

  (ii) Suggest formulae for the fragmentation ions responsible for the peaks at m/z values of 31 and 57.

  31 _____

  57 _____ [2]

121

7 Predict three fragmentation ions that you would expect to see in the mass spectrum of butan-1-ol and state the m/z value of each ion.

_____
_____
_____ [2]

8 After analytical analysis on an unbranched organic compound A the following results were obtained.

Percentage composition by mass	C 70.59% H 13.72 % O 15.69%
Infrared spectroscopy	Broad absorption at 3250 cm^{-1}
Mass spectrometry	Molecular ion peak at m/z = 102

Use this information to suggest all the possible structures for the unbranched compound A.

_____
_____
_____
_____
_____
_____
_____
_____
_____
_____ [6]

9 The structure of a carboxylic acid is shown below.

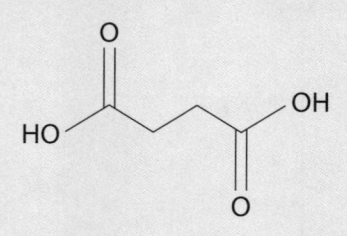

(a) Suggest the IUPAC name for this acid.

_____ [1]

(b) The mass spectrum of this acid is shown below.

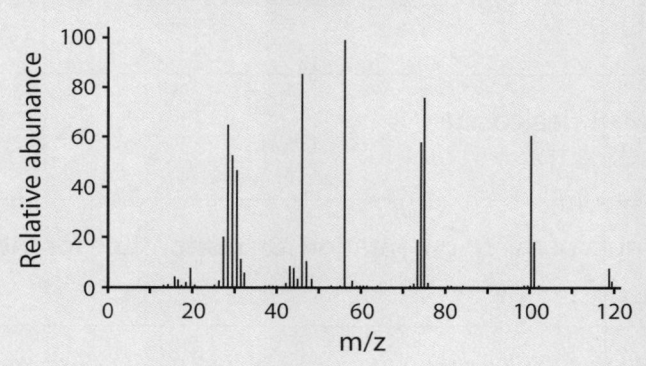

(i) Identify the base peak.

_____ [1]

(ii) State what is meant by the term **M+1 peak**.

_____

_____ [1]

(iii) State the m/z value of the M+1 peak.

_____ [1]

(iv) Suggest formulae of the fragmentation ions at the following m/z values.

45 _____

73 _____ [2]

10 The mass spectrum of an alcohol is shown below.

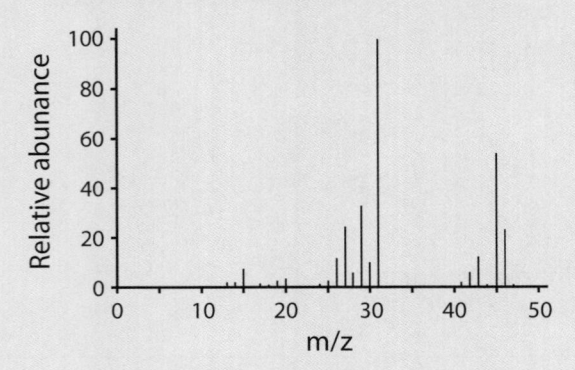

(a) State the relative molecular mass of the alcohol.

_____ [1]

(b) Explain why an M+1 peak occurs.

_____ [1]

(c) Suggest the formula of the fragmentation ion responsible for the base peak.

_____ [1]

(d) Suggest the formula of the fragmentation ions responsible for the peaks at m/z values of 15, 17 and 29.

15 _____

17 _____

29 _____ [3]

(e) Suggest the formula of the molecular ion.

_____ [1]

5.1 MASS SPECTROMETRY

11 The mass spectrum of two isomers of C_4H_9OH are shown below. One is 2-methyl propan-2-ol and the other is butan-1-ol.

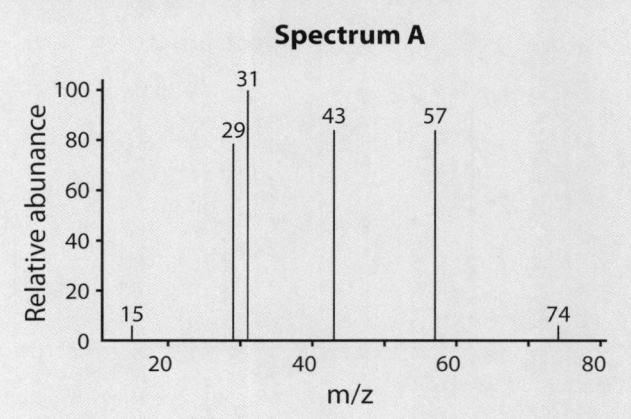

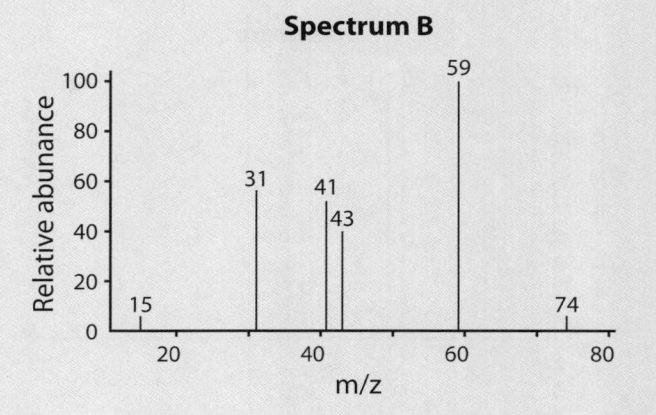

Spectrum A

Spectrum B

(a) Identify the ions responsible for the peaks at m/z values of 31 and 57 in spectrum A.

31 _____

57 _____ [2]

(b) Identify the ions responsible for the peaks at m/z values of 59 and 74 in spectrum B.

59 _____

74 _____ [2]

(c) Explain which mass spectrum (A or B) is that of butan-1-ol.

_____

_____ [2]

12 The mass spectrum of butan-1-ol and ethoxyethane, CH$_3$CH$_2$OCH$_2$CH$_3$, are shown below but not necessarily in that order.

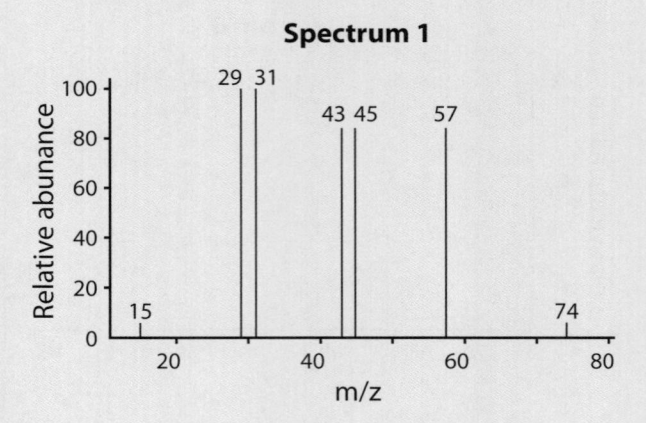

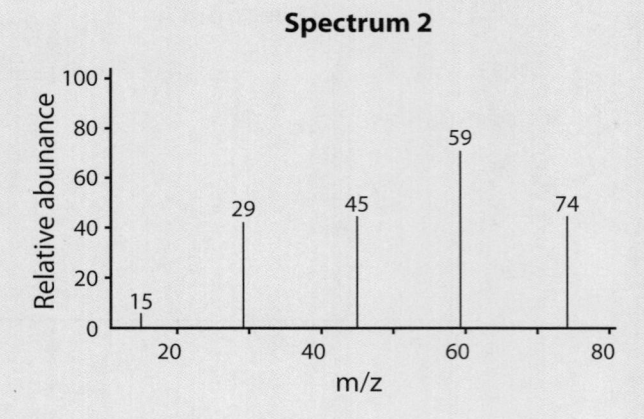

Spectrum 1

Spectrum 2

(a) Draw the structure of butan-1-ol and ethoxyethane showing all bonds.

butan-1-ol   ethoxyethane

[2]

(b) What is the m/z value of the base peak in spectrum 1?

_____[1]

(c) Identify the fragmentation ions giving rise to the peaks below in spectrum 1.

29 _____

45 _____

57 _____ [3]

(d) Use the fragmentation pattern in each spectrum to state which is ethyoxyethane. Explain your reasoning.

_____

_____ [2]

5.1 MASS SPECTROMETRY

13 Complete the table by giving the formula of fragmentation ions which occur in the mass spectrum of the different compounds shown. Provide two different possible fragmentation ions where "or" is indicated.

Compound	m/z	Fragmentation ion
butanone	29	
butanone	43	
ethyl methanoate	29	or
ethyl methanoate	45	or
ethyl methanoate	74	
$CH_3CH_2CONH_2$	57	
$CH_3CH_2CONH_2$	44	

[7]

14 2-chloropropane has two molecular ion peaks, one at m/z = 78 and the other at m/z = 80 in its mass spectrum.

(a) Explain why there are two molecular ion peaks.

_____

_____

_____ [2]

(b) Identify the fragmentation ions giving peaks at m/z = 35 and m/z = 63 in this spectrum.

35 _____

63 _____ [2]

5.2 Nuclear Magnetic Resonance Spectroscopy

1 Which one of the following factors determines the chemical shift in ^{1}H nuclear magnetic resonance spectroscopy?

    A   The chemical environment of hydrogen atoms

    B   The fragmentation of hydrogen atoms from the molecule

    C   The number of chemically equivalent hydrogen atoms

    D   The ratio of hydrogen atoms   [1]

2 An organic compound is shown below.

$$Cl-CH_2-\underset{\underset{CH_3}{|}}{\overset{\overset{CH_3}{|}}{C}}-CH_2-\overset{a}{CH_2}-Cl$$

(a) Give the IUPAC name for this compound.

_____ [1]

(b) Give the number of peaks in the low resolution ^{1}H nmr spectrum of this compound.

_____ [1]

(c) State and explain the splitting pattern for the peak due to the hydrogen labelled **a**, in the high resolution nmr spectrum of the compound.

_____

_____ [2]

(d) Explain the difference between a high resolution nmr spectrum and a low resolution nmr spectrum.

_____

_____ [2]

5.2 NUCLEAR MAGNETIC RESONANCE SPECTROSCOPY

3 The structure of hexane-2,5-dione, a toxic colourless liquid is shown below.

$$CH_3-\underset{\underset{O}{\|}}{C}-CH_2\overset{b}{-}CH_2-\underset{\underset{O}{\|}}{C}-CH_3$$

(a) Give the number of peaks in the low resolution ¹H nmr spectrum of this compound.

_____ [1]

(b) State and explain the splitting pattern for the peak due to the hydrogen labelled **b**, in the high resolution nmr spectrum of the compound.

_____

_____ [2]

(c) Explain the term **integration curve**.

_____

_____ [1]

(d) State the integration ratio for the hydrogen atoms in the chemically equivalent environments for hexane-2,5-dione.

_____ [1]

4 The substances CH₃CH₂CH₂CH₂OH and (CH₃)₃COH were dissolved in a suitable solvent and a ¹H nmr spectrum obtained for each.

(a) Identify a solvent which may be used.

_____ [1]

(b) State the number of peaks in the low resolution ¹H nmr spectrum of each substance.

CH₃CH₂CH₂CH₂OH _____

(CH₃)₃COH _____ [2]

(c) Name and draw the structure of the compound used as a standard in nmr.

_____ [2]

(d) State and explain two reasons why the compound named in (c) is used as a standard.

_____

_____ [2]

5 Describe and explain the appearance of the ¹H nmr spectrum of butenedioic acid. The skeletal formula of butenedioic acid is shown below.

_____

_____

_____

_____ [3]

6 The structure of compound A is shown below.

Compound A

On the axis shown below, sketch the expected ¹H nmr spectrum of compound A. Include the integration.

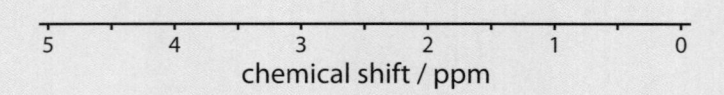

chemical shift / ppm

[2]

7 Which amine has three peaks in its proton nmr spectrum?

A   Methylamine

B   Trimethylamine

C   Diethylamine

D   Propylamine   [1]

8 Which one of the following shows the correct spin-spin splitting pattern in the ^{1}H nmr spectrum for propanone?

A   one singlet

B   singlet and triplet

C   two singlets

D   two triplets   [1]

9 The ^{1}H nmr spectrum of compound B is shown below.

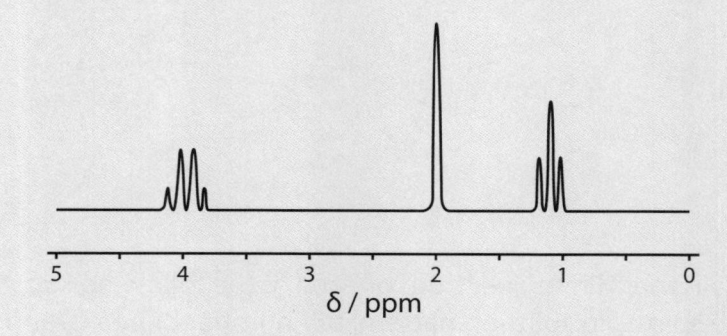

Which one of the following is B?

A   CH$_3$CH$_2$CH$_2$COOH

B   CH$_3$CH$_2$COOCH$_2$CH$_3$

C   CH$_3$COOCH$_2$CH$_3$

D   CH$_3$CH$_2$COOH   [1]

A2 2: ANALYTICAL, TRANSITION METALS, ELECTROCHEMISTRY AND ORGANIC NITROGEN CHEMISTRY

10 Nuclear magnetic resonance spectroscopy (nmr) is an important analytical technique used to determine the structure of different compounds.

  (a) Two isomers with the molecular formula $C_4H_8O_2$ have a triplet, a singlet and a quartet in their nmr spectrum. Draw the structures of the two isomers.

  [2]

  (b) Two isomers have the molecular formula $C_6H_{12}O_2$. They are esters and both have two singlets in their nmr spectrum with integration ratio 3:1. Draw two possible structures for the isomers.

  [2]

11 Describe and explain the ^1H nmr spectrum of each of the following substances. In your answer deduce the spin-spin splitting pattern, the number of peaks, and the integration.

  (a) ethanol

  _____
  _____
  _____
  _____ [3]

  (b) ethoxyethane $CH_3CH_2OCH_2CH_3$

  _____
  _____
  _____
  _____ [3]

12 The infrared spectrum and the ¹H nmr spectrum of a compound with a molecular formula $C_4H_8O_2$ are shown below.

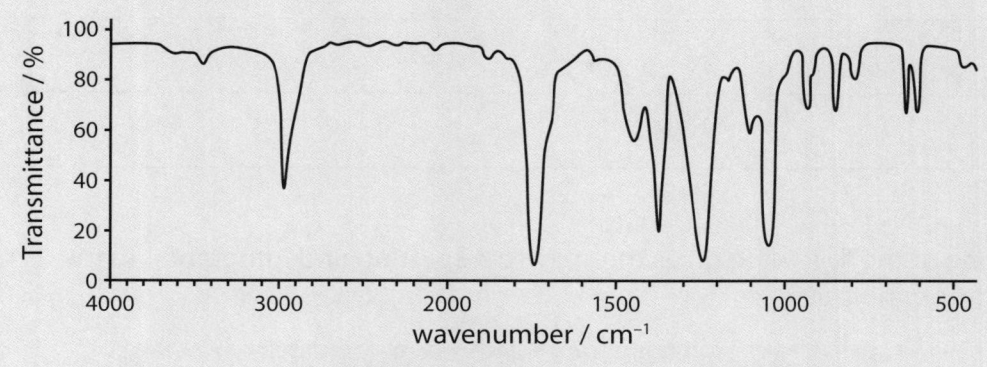

Infrared spectrum

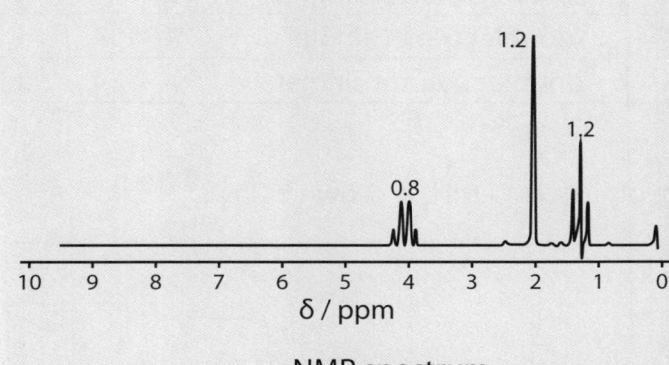

NMR spectrum

The relative integration values for the peaks are shown on the nmr spectrum.

Deduce the structure of the compound by analysing both spectra. Explain your deductions and draw the structural formula of the compound.

In this question you will be assessed on using your written communication skills including the use of specialist scientific terms.

_____

_____

_____

_____ [6]

13 Which one of the following gives the spin-spin splitting and integration in the ¹H nmr of 2-chloropropanoic acid?

	Spin-spin splitting	Integration		
A	triplet singlet singlet	3	1	1
B	doublet quartet singlet	3	1	1
C	doublet triplet singlet	3	4	1
D	doublet quartet singlet	2	4	1

14 The ¹H nmr spectrum of $CH_3CH_2CONH_2$ is shown below.

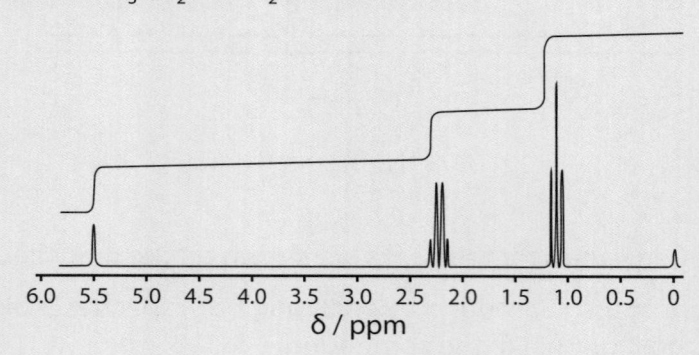

(a) State why there is a signal at δ = 0 ppm

_____ [1]

(b) Explain why the signal at δ = 1.15 ppm is a triplet.

_____

_____

_____ [2]

(c) Explain why the signal at δ = 2.25 ppm is a quartet.

_____

_____

_____ [2]

(d) Explain why the signal at δ = 5.5 ppm is at the highest chemical shift in the spectrum.

_____

_____ [1]

(e) Explain three ways in which the spectrum of the N-methylated compound, $CH_3CH_2CONHCH_3$, would differ from this spectrum.

_____

_____

_____ [2]

(f) What is meant by the term **triplet** in nmr spectroscopy?

_____

_____ [2]

5.3 Volumetric Analysis

Iodine-thiosulfate titrations

1   Which one of the following shows the correct sequence of colours for a titration of an acidified solution of potassium iodate(V) when excess potassium iodide (KI) is added and the solution is titrated with sodium thiosulfate (Na$_2$S$_2$O$_3$) using starch indicator near the end point?

	Colour change when excess KI added	Colour change as Na$_2$S$_2$O$_3$ added	Colour change at end point
A	Brown to colourless	Colourless to straw	Blue-black to colourless
B	Brown to colourless	Straw to colourless	Colourless to blue-black
C	Colourless to brown	Brown to straw	Blue-black to colourless
D	Colourless to brown	Straw to brown	Colourless to blue-black

2   A 25.0 cm³ sample of a solution of hydrogen peroxide was acidified using 1.50 mol dm⁻³ sulfuric acid. Excess solid potassium iodide was added and the solution was titrated against 0.245 mol dm⁻³ sodium thiosulfate solution. Starch indicator was added near the end point.

The results of several of these titrations are shown in the table below

	Titration 1	Titration 2	Titration 3	Titration 4
Final burette reading /cm³	18.5	36.3	17.1	35.0
Initial burette reading /cm³	0.0	18.5	0.0	17.1
Titre /cm³	18.5	17.8	17.1	17.9

(a) State the colour change observed at the end point.

_____ [1]

(b) Write an ionic equation for the reaction between hydrogen peroxide, sulfuric acid and potassium iodide.

_____ [2]

(c) Explain why the mean titre is quoted as 17.9 cm³ based on the titration results.

_____
_____
_____ [3]

(d) The reaction between thiosulfate ions and iodine is:

$$2S_2O_3^{2-} + I_2 \rightarrow S_4O_6^{2-} + 2I^-$$

Using the mean titre of 17.9 cm³ calculate the concentration of the hydrogen peroxide solution in mol dm⁻³. Give your answer to an appropriate number of significant figures.

_____

_____

_____

_____

_____ [4]

(e) Calculate the minimum volume of 1.50 mol dm⁻³ sulfuric acid required for each 25.0 cm³ sample of hydrogen peroxide solution based on the concentration calculated in (d).

_____

_____

_____ [3]

(f) Suggest why an excess of potassium iodide was added.

_____

_____

_____ [1]

3 A 250.0 cm³ sample of a solution of potassium iodate(V), KIO₃, of concentration 0.0540 M was prepared using a volumetric flask.

(a) Describe how this solution is prepared giving practical details.

In this question you will be assessed on using your written communication skills including the use of specialist scientific terms.

_____

_____

_____

_____

_____

_____

_____

_____

_____

_____ [6]

(b) 25.0 cm³ samples of the solution prepared in (a) were acidified using sulfuric acid and excess potassium iodide was added. The solution was then titrated using 0.445 M sodium thiosulfate solution. Starch indicator was added near the end point.

The overall equations for the reactions which occur are:

Reaction 1: $KIO_3 + 3H_2SO_4 + 5KI \rightarrow 3I_2 + 3H_2O + 3K_2SO_4$

Reaction 2: $2Na_2S_2O_3 + I_2 \rightarrow Na_2S_4O_6 + 2NaI$

(i) Write half equations for the oxidation and reduction processes in reaction 1.

oxidation _____

reduction _____ [2]

(ii) Write half equations for the oxidation and reduction processes in reaction 2.

oxidation _____

reduction _____ [2]

(iii) Explain why starch indicator was not added at the start of the titration.

_____

_____ [1]

(c) Calculate the volume, in cm³, of 0.445 M sodium thiosulfate solution which would be required to reach the end point. Give your answer to an appropriate number of significant figures.

_____

_____

_____

_____ [3]

4 24.8 g of hydrated sodium thiosulfate, $Na_2S_2O_3 \cdot xH_2O$, were dissolved in 1.00 dm³ of deionised water. This solution will be used to carry out a titration.

(a) 25.0 cm³ of 0.0150 mol dm⁻³ potassium iodate(V) solution were placed in a conical flask. The solution was acidified using sulfuric acid and excess potassium iodide was added.

(i) State the colour change which occurs in the solution and explain this colour change in terms of the reaction which occurs.

_____

_____

_____ [3]

(ii) Explain why sulfuric acid and not hydrochloric acid is used to acidify the solution in the conical flask.

_____

_____

_____ [2]

(b) The solution in (a) was titrated against the sodium thiosulfate solution and the mean titre was found to be 22.5 cm³.

(i) State the name of the indicator used in this titration and the colour change observed at the end point.

_____

_____ [2]

(ii) Write an ionic equation for the reaction between thiosulfate ions and iodine.

_____ [1]

(iii) Calculate the value of x in $Na_2S_2O_3 \cdot xH_2O$.

_____

_____

_____

_____

_____ [5]

A2 2: ANALYTICAL, TRANSITION METALS, ELECTROCHEMISTRY AND ORGANIC NITROGEN CHEMISTRY

5. A sample of 10.0 cm³ of a solution of hydrogen peroxide is diluted to 250.0 cm³ in a volumetric flask. A 25.0 cm³ portion of this solution is placed in a conical flask. Excess sulfuric acid and potassium iodide are added. The solution is titrated with 0.100 M sodium thiosulfate solution and the mean titre determined to be 12.5 cm³.

   Which one of the following is the molarity of the original hydrogen peroxide solution?

   A   0.0125 M

   B   0.0250 M

   C   0.625 M

   D   1.25 M   [1]

6. 2.44 g of an unknown iodate(V) salt, MIO_3, were dissolved in 250.0 cm³ of deionised water. 25.0 cm³ of this solution were placed in a conical flask and acidified. Excess potassium iodide was added and the solution titrated against 0.640 mol dm⁻³ sodium thiosulfate solution. The results of the titration are shown in the table below.

	Rough titration	Titration 1	Titration 2	Titration 3
Initial burette reading /cm³	00.00	12.70	24.20	31.00
Final burette reading /cm³	12.70		36.05	
Titre /cm³		11.50		11.60

   (a) Complete the blank values in the table.   [1]

   (b) Explain why a rough titration was carried out.

   _____

   _____

   _____ [1]

   (c) Using the same apparatus, suggest how the % uncertainty of the titre in titration 3 could be improved. Explain your answer.

   _____

   _____

   _____ [2]

5.3 VOLUMETRIC ANALYSIS

(d) Calculate the mean titre and give your answer to an appropriate number of significant figures. Justify which titration values you used.

_____

_____

_____ [2]

(e) Complete the statements below about the ratios in the reactions.

1 mol of IO_3^- produces _____ mol of I_2

1 mol of I_2 reacts with _____ mol of $S_2O_3^{2-}$

1 mol of IO_3^- requires _____ mol of $S_2O_3^{2-}$ [1]

(f) Calculate the relative formula mass of MIO_3 and suggest the identity of M.

_____

_____

_____

_____

_____ [5]

7 The percentage of copper in brass may be determined using an iodine-thiosulfate titration.

1. 0.245 g of brass were placed in a conical flask and reacted with concentrated nitric acid in a fume cupboard forming copper(II) ions, nitrogen(IV) oxide and water.

2. The mixture is warmed until the sample of brass has disappeared.

3. The nitric acid is removed by reaction with sodium carbonate solution.

4. Ethanoic acid is added to acidify the mixture.

5. Excess potassium iodide is added forming a precipitate of copper(I) iodide and iodine and the iodine formed is titrated with 0.125 mol dm^{-3} sodium thiosulfate solution using starch indicator. The end point is from blue-black to creamy-white.

(a) Write an ionic equation for the reaction of copper metal with hydrogen ions and nitrate(V) ions forming copper(II) ions, nitrogen(IV) oxide and water.

_____ [1]

(b) Nitric acid is removed in Step 3 because nitrate(V) ions can oxidise iodide to iodine in strongly acidic conditions. Write an ionic equation for the reaction of nitrate(V) ions with hydrogen ions and iodide ions forming iodine, nitrogen(IV) oxide and water.

_____ [1]

(c) Write an equation for the reaction of copper(II) ions with iodide ions forming copper(I) iodide and iodine.

_____ [1]

(d) Suggest why the colour change at the end point is from blue-black to creamy white.

_____
_____ [1]

(e) The equation for the reaction of iodine with sodium thiosulfate is:

$$2Na_2S_2O_3 + I_2 \rightarrow Na_2S_4O_6 + 2NaI$$

The mean titre is 14.2 cm³. Calculate the percentage of copper in the sample of brass.

_____
_____
_____
_____
_____ [5]

Manganate(VII) titrations

8 Which one of the following is correct for a titration of iron(II) ions against manganate(VII) ions?

	Indicator	**Colour change at end point**
A	methyl orange	yellow to red
B	no indicator	colourless to pink
C	phenolphthalein	pink to colourless
D	starch	blue-black to colourless

[1]

5.3 VOLUMETRIC ANALYSIS

9 A sample of 10.0 cm³ of a saturated solution of iron(II) sulfate at 20 °C is diluted to 250.0 cm³ using deionised water in a volumetric flask.

(a) Describe how the dilution is carried out.

_____

_____

_____

_____

_____ [5]

(b) Three 25.0 cm³ samples of the diluted solution were acidified with sulfuric acid and titrated against 0.0175 M potassium manganate(VII) solution. The titration results are shown in the table below.

	Initial burette reading / cm³	Final burette reading / cm³	Titre / cm³
Rough titration	0.0	22.5	
Accurate titration 1	22.5		21.6
Accurate titration 2		42.4	21.7

(i) Complete the missing values in the table. [1]

(ii) Calculate the mean titre.

_____ [1]

(iii) Calculate the molarity of the saturated solution of iron(II) sulfate. Give your answer to an appropriate number of significant figures.

_____

_____

_____

_____ [5]

(iv) Calculate the solubility of iron(II) sulfate, $FeSO_4$, in g per 100 g of water at 20 °C. Give your answer to an appropriate number of significant figures.

_____

_____ [2]

10 A sample of 2.75 g of impure iron(II) sulfate was dissolved in sulfuric acid and the volume made up to 250.0 cm³ using deionised water.

25.0 cm³ portions of the solution were acidified with sulfuric acid and titrated against 0.0150 mol dm⁻³ potassium manganate(VII) solution. The mean titre was found to be 17.85 cm³.

(a) Write an ionic equation for the reaction between iron(II) ions and manganate(VII) ions.

_____ [1]

(b) State the colour change at the end point.

_____ [1]

(c) Calculate the percentage of iron(II) sulfate in the sample.

_____

_____

_____

_____

_____

_____ [6]

11 Both iron(II) ions and ethanedioate ions react with manganate(VII) ions. Three half equations are given below.

Half equation 1: $Fe^{2+} \rightarrow Fe^{3+} + e^-$

Half equation 2: $MnO_4^- + 8H^+ + 5e^- \rightarrow Mn^{2+} + 4H_2O$

Half equation 3: $C_2O_4^{2-} \rightarrow 2CO_2 + 2e^-$

(a) Write an ionic equation for the reaction of iron(II) ions with manganate(VII) ions.

_____ [1]

(b) Write an ionic equation for the reaction of ethanedioate ions with manganate(VII) ions.

_____ [1]

(c) Write an overall equation for the reaction of iron(II) ethanedioate with manganate(VII) ions.

_____ [2]

(d) A sample of 25.0 cm³ of a solution of iron(II) ethanedioate, FeC$_2$O$_4$, was acidified with sulfuric acid and titrated against 0.00120 mol dm^{-3} potassium manganate(VII). 17.7 cm³ of the potassium manganate(VII) solution were required.

   (i) Explain why no indicator is required in this titration.

_____

_____ [2]

   (ii) Calculate the concentration of iron(II) ethanedioate, in mg dm^{-3}. Give your answer to an appropriate number of significant figures.

_____

_____

_____

_____ [4]

12 Hydrated ammonium iron(II) sulfate, (NH$_4$)$_2$Fe(SO$_4$)$_2$.6H$_2$O, contains water of crystallisation. The value of x decreases as the sample loses water of crystallisation to the atmosphere in a process called efflorescence.

An old sample of 19.6 g of hydrated ammonium iron(II) sulfate was dissolved in 1 dm³ of deionised water. 25.0 cm³ of this solution was acidified with sulfuric acid and titrated using 0.0130 mol dm^{-3} potassium manganate(VII) solution. The mean titre was 20.8 cm³.

(a) Calculate the value of x in this sample of hydrated ammonium iron(II) sulfate (NH$_4$)$_2$Fe(SO$_4$)$_2$.xH$_2$O.

_____

_____

_____

_____

_____ [5]

(b) Explain why the value you calculated in (a) is not a whole number.

_____

_____ [1]

(b) If the same process was carried out using the same mass of a fresh sample of hydrated ammonium iron(II) sulfate, $(NH_4)_2Fe(SO_4)_2.6H_2O$, calculate the volume of 0.0130 mol dm^{-3} potassium manganate(VII) solution which would be required. Give your answer to 3 significant figures.

_____

_____

_____

_____ [4]

Back titrations

13 A sample of 1.00 g of an impure metal was reacted with an excess of hydrochloric acid (50.0 cm³ of 2.50 mol dm^{-3}). The solution formed was transferred to a 1.00 dm³ volumetric flask and the volume made up using deionised water.

25.0 cm³ of this solution were titrated with 0.0750 M sodium hydroxide solution using phenolphthalein indicator.

This procedure was used for several metals. The impurities in the metals do not react with the acid.

(a) State the colour change of the indicator at the end point.

_____ [1]

(b) Write equations for the reactions of magnesium, aluminium, calcium and zinc with hydrochloric acid.

magnesium _____

aluminium _____

calcium _____

zinc _____ [4]

(c) Complete the table below. Use the space below the table for working out.

Metal	Mean titre / cm³	Percentage of metal in sample / %
Magnesium	19.9	
Aluminium	12.0	
Calcium		75.2
Zinc		86.7

_____
_____
_____
_____
_____
_____
_____
_____
_____ [6]

(d) Explain why the result for calcium may not be accurate.

_____
_____ [1]

14 Five indigestion tablets containing calcium carbonate were reacted with 50.0 cm³ of 1.50 mol dm⁻³ hydrochloric acid. The remaining solution was transferred to a volumetric flask and diluted to 250.0 cm³ using deionised water. Several 25.0 cm³ samples of this solution were titrated with 0.350 mol dm⁻³ sodium hydroxide solution using phenolphthalein indicator. The results of these titrations are given in the table below.

	Rough	Titration 1	Titration 2	Titration 3
Initial burette reading /cm³	0.0	14.2	28.0	30.0
Final burette reading /cm³	14.2	28.0	41.4	43.9
Titre /cm³	14.2	13.8	13.4	13.9

(a) Write two equations for the reactions which are occurring during this procedure.

_____

_____ [2]

(b) State the colour change observed at the end point.

_____ [1]

(c) Calculate the mean titre and justify your choices of titres.

_____

_____ [2]

(d) Calculate the mass, in mg, in one indigestion tablet. Give your answer to an appropriate number of significant figures.

_____

_____

_____

_____

_____ [5]

5.3 VOLUMETRIC ANALYSIS

15 A sample of 2.00 g of impure calcium carbonate was reacted with an excess of hydrochloric acid (50.0 cm³ of 1.00 M). The resulting solution was placed in a volumetric flask and diluted to 250.0 cm³. Several 25.0 cm³ portions of this solution were titrated against 0.100 M sodium hydroxide solution. The mean titre was 15.6 cm³. Calculate the percentage purity of the sample of calcium carbonate.

_____
_____
_____
_____
_____ [5]

16 2.25 g of an impure sample of barium hydroxide, $Ba(OH)_2$, were allowed to react with 50 cm³ of 0.750 mol dm⁻³ hydrochloric acid. The excess acid was titrated against 0.500 mol dm⁻³ sodium hydroxide solution. The mean titre was 27.95 cm³.

(a) Write an equation for the reaction of barium hydroxide with hydrochloric acid.

_____ [1]

(b) Calculate the percentage purity of the sample of barium hydroxide.

_____
_____
_____
_____
_____ [5]

17 A sample of 4.25 g of chromium metal was heated in oxygen and chromium(III) oxide formed. It was estimated that 95.2 % of the chromium was converted to chromium(III) oxide.

The heated sample was reacted with 100.0 cm³ of 3.50 mol dm⁻³ hydrochloric acid (an excess). Chromium metal does not react with hydrochloric acid. The resulting solution was filtered and diluted to 2.00 dm³ using deionised water. 25.0 cm³ of the resulting solution were titrated against 0.145 mol dm⁻³ sodium hydroxide solution using methyl orange.

(a) Write an equation for the formation of chromium(III) oxide from chromium.

_____ [1]

(b) Write an equation for the reaction of chromium(III) oxide with hydrochloric acid.

_____ [1]

(c) State the colour change observed in the solution when chromium(III) oxide reacts with hydrochloric acid.

_____ [1]

(d) Suggest why the solution was diluted to 2 dm³ rather than to 250 cm³.

_____

_____ [1]

(e) Calculate the volume of sodium hydroxide solution required based on the estimated percentage of chromium which reacted.

_____

_____

_____

_____

_____

_____

_____ [5]

(f) Calculate the percentage of chromium which reacts if the mean titre was 17.4 cm³.

_____

_____

_____

_____

_____

_____ [5]

5.4 Chromatography

1 Which of the following gases could **not** be used in gas-liquid chromatography as the mobile phase?

   A   argon

   B   neon

   C   nitrogen

   D   oxygen   [1]

2 The following 2-way paper chromatogram was developed for a mixture of amino acids. The spots are labelled W, X, Y and Z. Solvent 1 was a mixture of ethanol and water in a 4:1 ratio and solvent 2 was a mixture of propanone and hexane in a 1:2 ratio.

   Each small square on the graph is taken to be 1 mm by 1 mm.

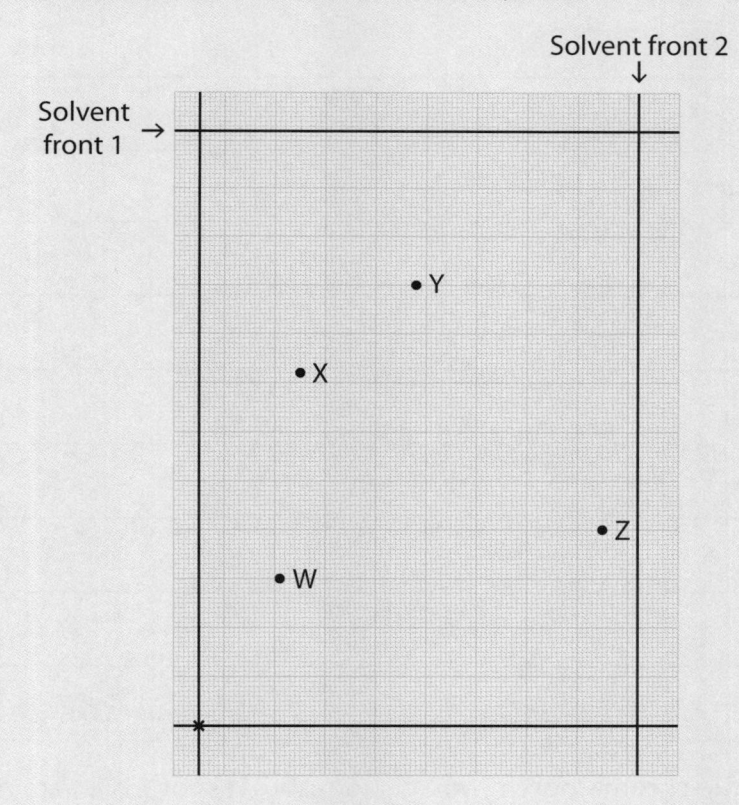

(a) Complete the table below. Give the R_f values to 3 significant figures.

Spot	Distance moved by spot in solvent 1 / mm	Distance moved by spot in solvent 2 / mm	R_f value solvent 1	R_f value solvent 2
W				
X				
Y				
Z				

[8]

(b) Explain how the 2-way chromatogram was set up and run.

In this question you will be assessed on using your written communication skills including the use of specialist scientific terms.

[6]

(c) Explain how the chromatogram would suggest that spot Y is an amino acid with a polar side chain and Z is an amino acid with a non-polar side chain.

[2]

5.4 CHROMATOGRAPHY

3 Complete the table below for the types of chromatography shown.

Chromatography	Mobile phase	Stationary phase
Gas-liquid chromatography		film of liquid on solid support
TLC		
Paper chromatography		water bonded to cellulose on the paper

[4]

4 A TLC chromatography experiment using a mixture of four dyes was carried out using a non-polar solvent. The following information was obtained.

Distance moved by solvent / mm	65
R_f value for dye A	0.615
R_f value for dye B	0.192
R_f value for dye C	0.885
R_f value for dye D	0.308

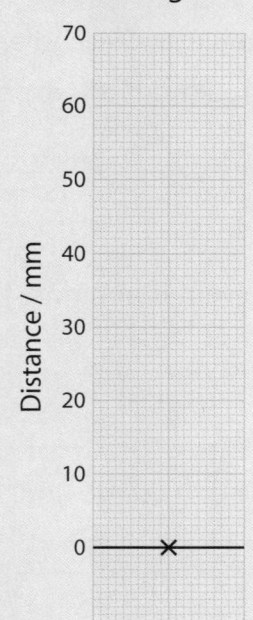

Chromatogram

(a) Explain why the origin line was drawn in pencil.

_____

_____ [1]

153

(b) Draw the solvent front line on the chromatogram and draw spots for the 4 dyes and label them A, B, C and D. [3]

(c) Explain which of the dyes (A, B, C or D) is the least polar.

_____

_____ [2]

5 The following procedure was used to carry out a thin-layer chromatography (TLC) experiment on a mixture of organic compounds using a volatile and toxic solvent to determine their R_f values.

1. Wear gloves throughout this experiment.
2. Draw a pencil line 1.5 cm from the bottom edge of the TLC plate. Draw a small cross in pencil at the centre of the line.
3. Use a capillary tube to make a small dot (no more than 1–2 mm in diameter) on the cross for the sample.
4. Add the solvent to a developing tank to a depth of no more than 1 cm.
5. Insert the TLC plate into the solvent.
6. Allow the solvent to rise up the plate until it is close to the top.
7. Remove the plate, mark the position of the solvent front and leave it to dry in a fume cupboard.
8. Place the plate under UV light and draw around each spot.

(a) Explain why gloves are worn throughout this experiment.

_____

_____ [1]

(b) Explain why the depth of the solvent is not more than 1 cm.

_____

_____ [1]

(c) Explain why the solvent is allowed to rise up until it is near the top of the plate.

_____

_____ [1]

(d) Why is the plate allowed to dry in a fume cupboard?

_____

_____ [1]

(e) Explain why it is necessary to place the plate under UV light.

_____

_____ [1]

(f) A diagram of the TLC chromatogram is shown below.

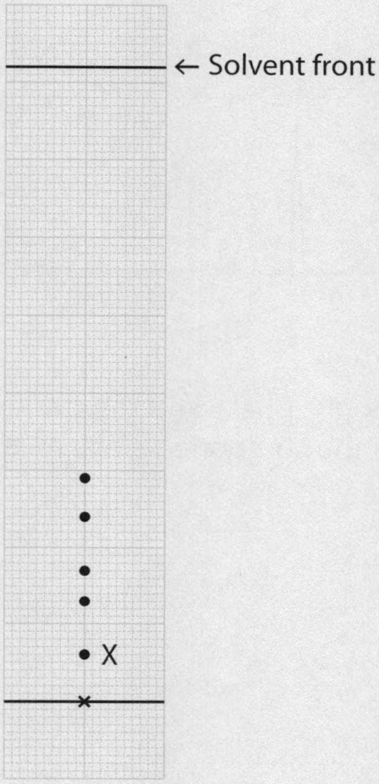

← Solvent front

•
•
•
•
• X
×

(i) Calculate the R$_f$ value of the spot labelled X.

_____

_____ [2]

(ii) Explain why the component responsible for spot X may be more polar than the other components in the mixture.

_____

_____ [2]

(iii) Explain why the experiment would be carried out again with a different solvent.

_____

_____ [1]

6 A mixture of organic compounds was analysed by gas-liquid chromatography (GLC) and the trace is shown below. Four component peaks were found and are labelled K, L, M and N together with their relative integration values for the areas under the peaks.

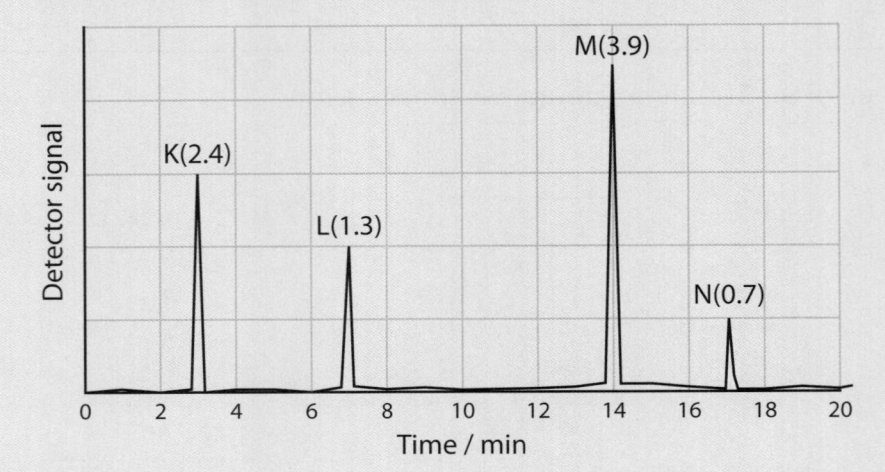

(a) Describe, in terms of the mobile phase and the stationary phase, how gas-liquid chromatography separates the components of a mixture.

_____

_____

_____

_____

_____ [4]

(b) Explain why it is not possible to state that the mixture used contained only 4 compounds.

_____

_____ [1]

(c) Suggest how the use of mass spectrometry combined with GLC could help solve the issue identified in (b).

_____

_____

_____ [2]

(d) Calculate the percentage of component N in the mixture.

_____

_____

_____ [2]

7 The R_f values for 5 amino acids analysed by TLC are shown in the table below. A non-polar solvent was used (ether/hexane/ethanol in a 3:3:1). The developing agent ninhydrin was sprayed on the TLC plate to locate the spots.

Amino acid	Code	Formula	R_f value (TLC silica support)
Lysine	lys	$H_2NCH(COOH)(CH_2)_4NH_2$	0.165
Cysteine	cys	$H_2NCH(COOH)CH_2SH$	0.372
Phenylalanine	phe	$H_2NCH(COOH)CH_2C_6H_5$	0.624
Tyrosine	tyr	$H_2NCH(COOH)CH_2C_6H_4OH$	0.551
Valine	val	$H_2NCH(COOH)CH(CH_3)_2$	0.440

(a) State what is the mobile phase and what is the stationary phase in this TLC experiment.

Mobile phase: _____

Stationary phase: _____ [2]

(b) Explain why it is necessary to spray the plate with ninhydrin to locate the spots.

_____

_____ [1]

(c) State another method or technique which may be used to locate the spots.

_____

_____ [1]

(d) Describe how R_f values are determined.

_____

_____

_____

_____ [4]

(e) On the TLC chromatogram below, draw small spots for the amino acids based on their R_f values and label them with the code. The origin and solvent front are marked.

⟵ Solvent front

✗

[3]

(f) Suggest why lysine has the lowest R_f value and phenylalanine the highest in this experiment.

_____

_____

_____ [3]

5.5 Transition Metals

General properties of transition metals and complexes

1. Which one of the following is the electronic configuration of a Cr^{3+} ion?

    A   [Ar] $3d^2\ 4s^1$

    B   [Ar] $3d^3$

    C   [Ar] $3d^1\ 4s^2$

    D   [Ar] $3d^1\ 4s^1$ [1]

2. In which one of the following complexes is iron **not** in the +2 oxidation state?

    A $[Fe(H_2O)_5(OH)]^{2+}$

    B $[Fe(C_2O_4)_2(H_2O)_2]^{2-}$

    C $[Fe(edta)]^{2-}$

    D $[FeCl_4]^{2-}$ [1]

3. Which one of the following is **not** correct?

	Complex	Oxidation state of transition metal	Shape of complex	Co-ordination number of complex
A	$[Ag(NH_3)_2]^+$	+1	linear	2
B	$[Cu(NH_3)_4(H_2O)_2]^{2+}$	+2	octahedral	6
C	$[Pt(NH_3)_2Cl_2]$	0	square planar	4
D	$[CoCl_4]^-$	+3	tetrahedral	4

[1]

4. Which one of the following **cannot** act as a ligand?

    A   hydride ions

    B   hydrogen molecules

    C   water molecules

    D   hydroxide ions [1]

5 The table below shows 5 different complexes.

A	B	C	D	E
$[Ni(H_2O)_6]^{2+}$	$[Ni(H_2NCH_2CH_2NH_2)_3]^{2+}$	$[Ni(edta)]^{2-}$	$[Co(NH_3)_4Cl_2]^+$	$[Ni(NH_3)_6]^{2+}$

(a) Draw and name the shape of complex A.

_____ [2]

(b) Write an equation to show how complex B is converted into complex C.

_____ [1]

(c) Name the ligand in complex B.

_____ [1]

(d) State the co-ordination number in complex C.

_____ [1]

(e) What is the oxidation state of cobalt in complex D?

_____ [1]

(f) Complex E is converted into complex B as shown in the equation below.

$$[Ni(NH_3)_6]^{2+} + 3H_2NCH_2CH_2NH_2 \rightarrow [Ni(H_2NCH_2CH_2NH_2)_3]^{2+} + 6NH_3$$

(i) Explain why the enthalpy change for this reaction is approximately zero.

_____

_____ [2]

(ii) Explain why the reaction occurs despite having an enthalpy change of almost zero.

_____

_____ [2]

(g) Complex D exhibits cis trans (E-Z) isomerism. The cis isomer is shown below.

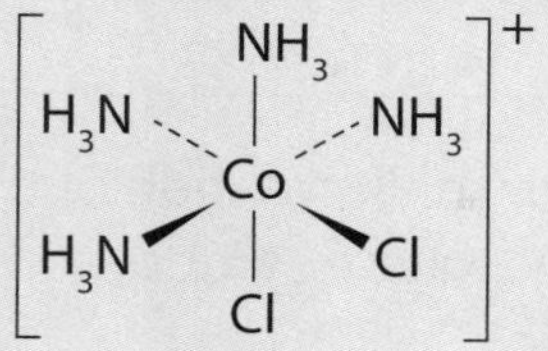

(i) Explain why the structure shown above is the cis isomer.

_____

_____ [1]

(ii) Draw the structure of the trans isomer.

[1]

6 The following reaction occurs when hydrochloric acid is added to a solution containing nickel(II) sulfate.

$[Ni(H_2O)_6]^{2+} + 4Cl^- \rightarrow [NiCl_4]^{2-} + 6H_2O$

(a) State and explain the change of co-ordination number during this reaction.

_____

_____

_____ [2]

(b) 3.30 g of hydrated nickel(II) sulfate, $NiSO_4.6H_2O$, were dissolved in 100 cm³ of deionised water. To a 10 cm³ sample of this solution, 10 cm³ of 1.5 mol dm⁻³ hydrochloric acid were added (an excess). An excess of 0.25 mol dm⁻³ silver nitrate solution was added to remove chloride ions which had not reacted. A precipitate was formed.

(i) What would be observed when nickel(II) sulfate was dissolved in water?

_____ [1]

(ii) Write an equation for the formation of $[Ni(H_2O)_6]^{2+}$ from nickel(II) sulfate.

_____ [1]

(iii) Write an ionic equation for the formation of the precipitate when silver nitrate solution was added. Include state symbols.

_____ [2]

(iv) Calculate the minimum volume of silver nitrate solution required to react with the chloride ions which did not react with the nickel complex. Give your answer to an appropriate number of significant figures.

_____

_____

_____

_____ [5]

(v) State the colour of the precipitate obtained when excess silver nitrate solution was added and calculate the mass, in mg, of precipitate obtained. Give your answer to an appropriate number of significant figures.

_____

_____

_____ [3]

7 The glycinate ion, $H_2NCH_2COO^-$ can act as bidentate ligand.

(a) What is meant by a **bidentate ligand**?

_____

_____ [2]

(b) State the IUPAC name for the glycinate ion.

_____ [1]

(c) Write the formula of the complex formed between copper(II) ions and glycinate ions which has a co-ordination number of 6.

_____ [1]

(d) Draw the structural of the complex described in (c). Show clearly the shape of the complex.

[1]

8 Complex A may be converted into complex B.

$$\left[\begin{array}{c} \text{H}_2\text{O} \cdots \overset{\text{OH}_2}{\underset{\text{OH}_2}{\overset{|}{\text{Fe}}}} \cdots \text{OH}_2 \\ \text{H}_2\text{O} \end{array}\right]^{3+}$$

Complex A

$$\left[\begin{array}{c} \text{Complex B structure with oxalate ligands and Fe} \end{array}\right]^{-}$$

Complex B

(a) Write an equation for this reaction.

_____ [1]

(b) Name the ligand which replaces the water molecules.

_____ [1]

(c) Suggest why the ligand replacement is only partial.

_____

_____ [1]

(d) Explain why the enthalpy change for the reaction is approximately zero and why the reaction occurs despite this.

_____

_____

_____

_____ [4]

(e) State the co-ordination number of complex B.

_____ [1]

9 The following reactions occur.

Reaction 1: $[Cu(H_2O)_6]^{2+} + 4NH_3 \rightleftharpoons [Cu(NH_3)_4(H_2O)_2]^{2+} + 4H_2O$

Reaction 2: $[Cu(H_2O)_6]^{2+} + 4Cl^- \rightleftharpoons [CuCl_4]^{2-} + 6H_2O$

(a) Name the complex $[Cu(H_2O)_6]^{2+}$.

_____ [1]

(b) Suggest, in terms of the availability of the lone pair of electrons, why ammonia replaces water ligands.

_____

_____

_____ [2]

(c) The Cl^- ligand is a weaker ligand than H_2O. Suggest why concentrated hydrochloric acid is used in reaction 2.

_____

_____

_____ [2]

(d) What would be observed in reaction 1?

_____

_____ [2]

(e) Complete the table below for the complexes shown.

Complex	Shape	Bond angle / °	Co-ordination number
$[Cu(H_2O)_6]^{2+}$			
$[CuCl_4]^{2-}$			

[2]

10 Redox and disproportionation reactions occur with transition metal complexes.

(a) What is meant by a transition metal?

_____

_____ [1]

(b) What is a complex?

_____

_____ [2]

(c) The following redox reaction occurs.

$[Fe(CN)_6]^{4-} + [IrCl_6]^{2-} \rightarrow [Fe(CN)_6]^{3-} + [IrCl_6]^{3-}$

$[IrCl_6]^{2-}$ is the hexachloroiridate(IV) ion.

(i) Explain why this reaction is a redox reaction.

_____

_____

_____

_____ [3]

(ii) Write the formula of potassium hexachloroiridate(IV).

_____ [1]

(iii) Suggest the name of the $[IrCl_6]^{3-}$ ion.

_____ [1]

(d) The following reaction occurs.

$$3[AuCl_2]^- \rightarrow 2Au + [AuCl_4]^- + 2Cl^-$$

(i) Suggest the shape of the $[AuCl_2]^-$ and $[AlCl_4]^-$ complexes.

$[AuCl_2]^-$ _____

$[AuCl_4]^-$ _____ [2]

(ii) Explain why this reaction is a disproportionation reaction.

_____

_____

_____

_____ [3]

11 Potassium tetrachloropalladate(II), $K_2[PdCl_4]$ and potassium tetrachloronickelate(II), $K_2[NiCl_4]$ are compounds containing complex ions.

(a) Write an equation for the formation of potassium tetrachloropalladate(II) from potassium chloride and palladium(II) chloride.

_____ [1]

(b) The tetrachloropalladate(II) ion is square planar whereas the tetrachloronickelate(II) ions is tetrahedral.
Draw the shape of these ions and indicate the bond angle.

tetrachloropalladate(II) ion   tetrachloronickelate(II) ion

[2]

(c) Tetrachloronickelate(II) ions react with 1,2-diaminoethane and also with edta^{4-}.

(i) Write an equation for the reaction of tetrachloronickelate(II) ions with three 1,2-diaminoethane molecules.

_____ [2]

(ii) Write an equation for the reaction of tetrachloronickelate(II) ions with one edta^{4-} ion.

_____ [2]

(iii) Complete the table below by placing a tick (✓) in the correct column for each ligand.

Ligand	Monodentate	Bidentate	Hexadentate
CN$^-$			
H$_2$O			
Cl$^-$			
1,2-diaminoethane			
edta^{4-}			

[2]

Colour of transition metal ions and their identification

12 Which one of the following would be observed when ammonia solution is added to a solution of cobalt(II) chloride?

	Colour of precipitate formed	Colour of solution formed when precipitate dissolves in excess ammonia solution
A	blue	pink
B	blue	yellow
C	pink	blue
D	pink	yellow

[1]

13 A precipitate forms when a few drops of solution Y is added to solution X. In which one of the following would the precipitate **not** be soluble in an excess of solution Y?

	X	Y
A	nickel(II) sulfate	ammonia
B	copper(II) sulfate	ammonia
C	chromium(III) sulfate	sodium hydroxide
D	iron(II) sulfate	sodium hydroxide

[1]

14 You are provided with samples of hydrated nickel(II) chloride and hydrated chromium(III) chloride. Both samples are green crystalline solids.

(a) Write the formula for chromium(III) chloride-6-water.

_____ [1]

(b) Describe how you would test the sample of hydrated nickel(II) chloride for chloride ions.

_____

_____

_____

_____ [4]

(c) Describe how you would carry out qualitative tests to positively identify the transition metal ions present in both of the solids using sodium hydroxide solution and aqueous ammonia. Write equations for any reactions which occur.

In this question you will be assessed on using your written communication skills including the use of specialist scientific terms.

_____

_____

_____

_____

_____

_____

_____

_____

_____

_____

_____ [6]

(d) Chromium(III) chloride dissolves in water to form hexaaquachromium(III) ions and chloride ions in solution. Some ligand replacement occurs as shown by the equation below. Chloride ions which act as ligands are not precipitated when silver nitrate solution is added.

$$[Cr(H_2O)_6]^{3+} + Cl^- \rightarrow [Cr(H_2O)_5Cl]^{2+} + H_2O$$

If 450 mg of hydrated chromium(III) chloride were dissolved in deionised water and the reaction above occurred, calculate the mass of silver chloride formed, in mg, if an excess of silver nitrate solution was added. Give your answer to 3 significant figures.

_____

_____

_____

_____ [3]

15 The following tests were carried out on two transition metal compounds labelled **A** and **B**.

	Test	Observations
1	Dissolve a sample of **A** in deionised water.	Green solution formed.
2	To the solution of **A**, add ammonia solution dropwise and then until it is in excess.	Green precipitate which is soluble in excess ammonia solution forming a blue solution.
3	To the solution of **A**, add dilute nitric acid followed by barium chloride solution.	White precipitate formed.
4	Dissolve a sample of **B** in deionised water.	Yellow solution formed.
5	To the solution of **B**, add sodium hydroxide solution and then until it is in excess.	Brown precipitate formed which is not soluble in excess sodium hydroxide solution.
6	To the solution of **B**, add dilute nitric acid and silver nitrate solution followed by dilute ammonia solution.	White precipitate which is soluble in dilute ammonia solution.

(a) Identify **A** and **B**.

A _____

B _____ [2]

A2 2: ANALYTICAL, TRANSITION METALS, ELECTROCHEMISTRY AND ORGANIC NITROGEN CHEMISTRY

   (b) Write the formulae of the following compounds or complexes.

      (i) Green precipitate formed in Test 2.

_____ [1]

      (ii) Blue complex in the solution in Test 2 when excess ammonia solution was added.

_____ [1]

      (iii) White precipitate formed in Test 3.

_____ [1]

      (iv) Yellow complex in the solution in Test 4.

_____ [1]

      (v) Brown precipitate formed in Test 5.

_____ [1]

      (vi) White precipitate formed in Test 6.

_____ [1]

      (vii) Complex formed when dilute ammonia solution is added in Test 6.

_____ [1]

   (c) Write an ionic equation for the formation of the brown precipitate in Test 5.

_____ [1]

   (d) What would be observed if sodium hydroxide solution was was added until it was in excess to the solution of **A**?

_____

_____ [2]

16 Which one of the following is the correct colour of the compound and oxidation state of the transition metal?

	chromium compound	colour	oxidation state
A	K_2CrO_4	orange	+6
B	$K_2Cr_2O_7$	yellow	+7
C	$KMnO_4$	purple	+7
D	$MnSO_4$	pink	+4

[1]

5.5 TRANSITION METALS

17 The table below shows the observations for the reactions of transition metal ions with ammonia solution.

Transition metal ion	Observation on addition of a few drops of ammonia solution	Observation on addition of excess ammonia solution
Mn^{2+}	white ppt which slowly changes to brown	ppt is insoluble
Fe^{2+}	green ppt	
Fe^{3+}		
Cr^{3+}		
Ni^{2+}		
Co^{2+}		
Cu^{2+}	blue ppt	ppt is soluble forming a dark blue solution

(a) Complete the table. [4]

(b) Write an ionic equation for the formation of the blue precipitate for Cu^{2+} ions.

_____ [1]

(c) Write an equation for the dissolution of the blue precipitate for Cu^{2+} ions forming the dark blue solution.

_____ [2]

(d) The white precipitate of manganese(II) hydroxide is oxidised by oxygen in the air to form brown hydrated manganese(III) oxide. Write an equation for this reaction.

_____ [2]

(e) Chromium(III) hydroxide may be oxidised to chromate(VI) ions using hydrogen peroxide. Chromate(VI) ions may be converted to dichromate(VI) ions by the addition of hydrogen ions (ethanoic acid).

  (i) State the colour changes which occur.

  _____

  _____ [2]

171

(ii) Write a half equation for the oxidation of chromium(III) hydroxide to chromate(VI) ions.

_____ [2]

(iii) The reduction of hydrogen peroxide is given by the half equation:

$$H_2O_2 + 2H^+ + 2e^- \rightarrow 2H_2O$$

Write an ionic equation for the oxidation of chromium(III) hydroxide to chromate(VI) using hydrogen peroxide.

_____ [2]

(iv) Write an ionic equation for the conversion of chromate(VI) into dichromate(VI).

_____ [1]

Vanadium chemistry

18 Using the electrode potentials below, choose the reducing agent capable of reducing vanadium from the +5 to the +4 oxidation state but no further.

Half-equation	$E^\ominus$ / V
$VO_2^+(aq) + 2H^+(aq) + e^- \rightarrow VO^{2+}(aq) + H_2O(l)$	+1.00
$VO^{2+}(aq) + 2H^+(aq) + e^- \rightarrow V^{3+}(aq) + H_2O(l)$	+0.32
$V^{3+}(aq) + e^- \rightarrow V^{2+}(aq)$	−0.26
$Ag^+(aq) + e^- \rightarrow Ag(s)$	+0.80
$Fe^{2+}(aq) + 2e^- \rightarrow Fe(s)$	−0.44
$Sn^{2+}(aq) + 2e^- \rightarrow Sn(s)$	+0.14
$SO_4^{2-}(aq) + 4H^+(aq) + 2e^- \rightarrow SO_2(g) + 2H_2O(l)$	+0.17

A   Ag(s)

B   Fe(s)

C   Sn(s)

D $SO_2(g)$

[1]

5.5 TRANSITION METALS

19 Vanadium shows a variety of oxidation states.

(a) Complete the table below for the oxidation states of vanadium and colours of the vanadium compounds.

Vanadium compound	Oxidation state of vanadium	Colour of compound
NH$_4$VO$_3$		
VSO$_4$		
VCl$_3$		
VO(NO$_3$)$_2$		

[4]

(b) Solid NH$_4$VO$_3$ is added to an excess of hydrochloric acid. The solution contains VO$_2$Cl.

(i) Write an ionic equation for the reaction which occurs.

_____ [1]

(ii) Explain why this is not a redox reaction.

_____
_____ [1]

(c) VO$_2$Cl may be reduced to VCl$_2$ by zinc in the presence of hydrochloric acid.

(i) Write an equation for the conversion of VO$_2$Cl to VCl$_2$.

_____ [2]

(ii) State the initial and final colours observed during this reaction.

initial colour: _____

final colour: _____ [1]

20 Vanadium(V) oxide acts as a catalyst in the Contact process for the production of sulfuric acid. The equations below show how it catalyses the oxidation of sulfur dioxide to sulfur trioxide.

SO$_2$(g) + V$_2$O$_5$(s) → SO$_3$(g) + V$_2$O$_4$(s)

V$_2$O$_4$(s) + ½O$_2$(g) → V$_2$O$_5$(s)

(a) Write an overall equation for the oxidation of sulfur dioxide to sulfur trioxide.

_____ [1]

(b) State the colour of vanadium(V) oxide.

_____ [1]

(c) Explain why vanadium(V) oxide is referred to as a heterogeneous catalyst in this process.

_____

_____

_____ [2]

(d) Vanadium(V) oxide reacts with nitric acid forming dioxovanadium(V) nitrate and water and also with an excess of sodium hydroxide forming sodium orthovanadate(V), Na_3VO_4 and water.

(i) Write an equation for the reaction of vanadium(V) oxide with nitric acid.

_____ [2]

(ii) Write an equation for the reaction of vanadium(V) oxide with excess sodium hydroxide.

_____ [2]

(iii) The orthovanadate(V) ions may be reduced to oxovanadium(IV) by the amino acid cysteine. Write a half equation for this reduction of orthdovanadate(V) to oxovanadium(IV).

_____ [1]

(iv) Explain why a green colour is observed during this reduction.

_____

_____ [2]

(v) Sodium orthovanadate(V) reacts with iron in presence of sulfuric acid forming sodium sulfate, iron(II) sulfate, vanadium(II) sulfate and water. Write an equation for this reaction.

_____ [2]

(v) State the colours of the following vanadium compounds.

VO_2Cl _____

$VOSO_4$ _____

VSO_4 _____

$V_2(SO_4)_3$ _____ [2]

21 Some standard electrode potentials below are given below.

	$E^⦵$ / V
$VO_2^+(aq) + 2H^+(aq) + e^- \rightleftharpoons VO^{2+}(aq) + H_2O(l)$	+1.00
$VO^{2+}(aq) + 2H^+(aq) + e^- \rightleftharpoons V^{3+}(aq) + H_2O(l)$	+0.32
$V^{3+}(aq) + e^- \rightleftharpoons V^{2+}(aq)$	−0.26
$I_2(aq) + 2e^- \rightleftharpoons 2I^-(aq)$	+0.54
$Br_2(aq) + 2e^- \rightleftharpoons 2Br^-(aq)$	+1.09
$Ag^+(aq) + e^- \rightleftharpoons Ag(s)$	+0.80
$Pb^{2+}(aq) + 2e^- \rightleftharpoons Pb(s)$	−0.13

(a) (i) Choose an oxidising agent which can oxidise vanadium from +2 to +3 but no further. Explain your answer using the electrode potential above.

_____

_____

_____ [3]

(ii) What colour change would be observed when vanadium is oxidised from the +2 to +3 oxidation state?

_____ [1]

(iii) Write an ionic equation for the reaction which occurs using the oxidising agent you chose in (a)(i).

_____ [1]

(b) Chlorine will oxidise vanadium from the +2 to the +5 oxidation state.

(i) Write an ionic equation for this reaction.

_____ [2]

(ii) What colour would be observed at the end of the reaction?

_____ [1]

(iii) Calculate the EMF values for each step in the oxidation using chlorine given the following electrode potential of chlorine.

$Cl_2(aq) + 2e^- \rightarrow 2Cl^-$ $E^\ominus = +1.36$ V

+2 to +3 _____

_____

+3 to +4 _____

_____

+4 to +5 _____

_____ [3]

5.6 Electrode Potentials

Electrochemical cells

1. Some electrode potentials are given below.

Electrode half equation	$E^\ominus$ / V
$Zn^{2+} + 2e^- \rightleftharpoons Zn$	−0.76
$Pb^{2+} + 2e^- \rightleftharpoons Pb$	−0.13
$Cu^{2+} + 2e^- \rightleftharpoons Cu$	+0.34
$Ag^+ + e^- \rightleftharpoons Ag$	+0.80

   Which one of the following is the strongest reducing agent?

   A Ag^+

   B   Cu

   C Pb^{2+}

   D   Zn   [1]

2. Which one of the following is the EMF for the cell $Cu^+|Cu^{2+}||Cu^+|Cu$ given the electrode potentials below?

   $Cu^{2+} + e^- \rightleftharpoons Cu^+$ $E^\ominus = +0.15$ V

   $Cu^+ + e^- \rightleftharpoons Cu$ $E^\ominus = +0.52$ V

   A   −0.67 V

   B   −0.37 V

   C   +0.37 V

   D   +0.67 V   [1]

3. The EMF of the cell below is +0.03 V

   $Fe^{2+}(aq)|Fe^{3+}(aq)||Ag^+(aq)|Ag(s)$

   and the electrode potential for the silver electrode is:

   $Ag^+(aq) + e^- \rightleftharpoons Ag(s)$ $E^\ominus = +0.80$ V

   Which one of the following is the electrode potential for the half equation below?

   $Fe^{3+}(aq) + e^- \rightleftharpoons Fe^{2+}(aq)$

   A   −0.77 V

   B   −0.83 V

   C   +0.77 V

   D   +0.83 V   [1]

4 Some standard electrode potentials are given below.

Half-equation	$E^\ominus$ / V
$Cl_2(g) + 2e^- \rightleftharpoons 2Cl^-(aq)$	+1.36
$Br_2(aq) + 2e^- \rightleftharpoons 2Br^-(aq)$	+1.09
$I_2(aq) + 2e^- \rightleftharpoons 2I^-(aq)$	+0.54
$2H^+(aq) + 2e^- \rightleftharpoons H_2(g)$	0.00
$Fe^{3+}(aq) + e^- \rightleftharpoons Fe^{2+}(aq)$	+0.77
$Fe^{2+}(aq) + 2e^- \rightleftharpoons Fe(s)$	−0.44
$O_2(g) + 4H^+(aq) + 4e^- \rightleftharpoons 2H_2O(l)$	+1.23

(a) Draw a labelled diagram to show how the Fe^{3+}/Fe^{2+} standard electrode potential is measured.

[6]

(b) Chlorine will oxidise iron to form iron(III) chloride.

(i) Write an overall equation for this reaction.

_____ [1]

(ii) Explain why chlorine oxidises iron to iron(III) chloride using the standard electrode potentials in the table.

_____

_____

_____ [2]

(iii) Predict the product formed when bromine reacts with iron and when iodine reacts with iron. Explain your answer using the standard electrode potentials.

_____

_____

_____

_____

_____

_____

_____ [6]

(c) The equation below shows a reaction of chlorine with water:

$$2Cl_2(g) + 2H_2O(l) \rightarrow O_2(g) + 4HCl(aq)$$

(i) Explain, using oxidation numbers, why this is a redox reaction.

_____

_____

_____ [3]

(ii) Using the information in the table at the start of the question, write the conventional representation for the cell for this reaction and calculate the EMF.

_____

_____

_____ [3]

(iii) Explain why this reaction may cause an issue with water treatment.

_____

_____ [2]

5 The following cell was set up. The amount of copper in the electrodes is much greater than the amount of Cu²⁺ ions in either of the solutions.

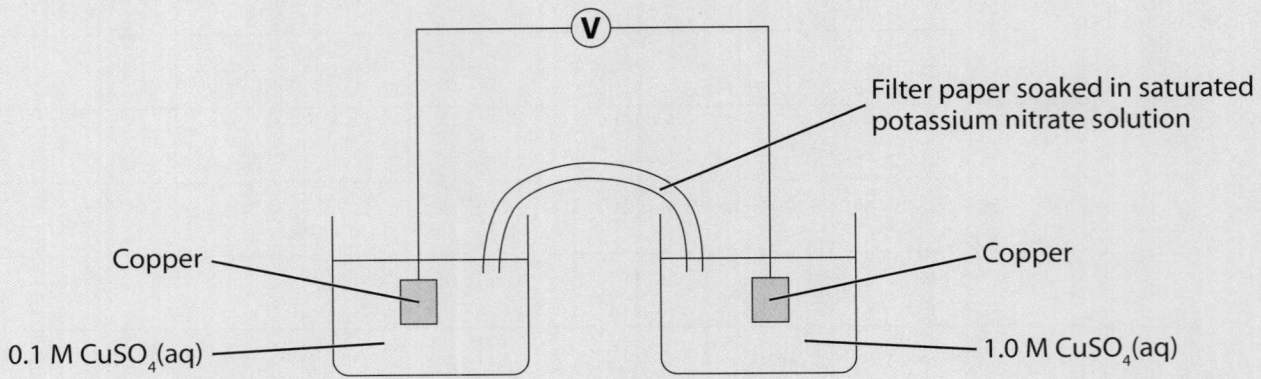

(a) What name is given to the filter paper soaked in saturated potassium nitrate solution and what is its function?

_____

_____

_____ [2]

(b) Suggest why a saturated solution of potassium chloride should not be used in place of the potassium nitrate.

_____

_____

_____ [1]

(c) If the voltmeter was replaced with an ammeter, electrons flow from left to right. Explain this in terms of the processes occurring in each half cell.

_____

_____

_____

_____ [4]

(d) Write the conventional cell representation of this cell.

_____ [2]

(e) Explain why the current would eventually be zero.

_____

_____ [1]

(f) The EMF of the cell is +0.03 V. The standard electrode potential for a Cu^{2+}/Cu electrode is +0.34 V.

(i) What would the EMF of the cell be if the concentrations of the solutions were both 1.0 M?

_____ [1]

(ii) Calculate the electrode potential of the cell containing 0.1 M $CuSO_4$ solution.

_____

_____ [2]

6 An electrochemical cell was set up using the two half cells represented by the equations below.

$Cu^{2+}(aq) + 2e^- \rightleftharpoons Cu(s)$ $E^\ominus = +0.34$ V

$Fe^{3+}(aq) + e^- \rightleftharpoons Fe^{2+}(aq)$ $E^\ominus = +0.77$ V

The diagram below represents this cell.

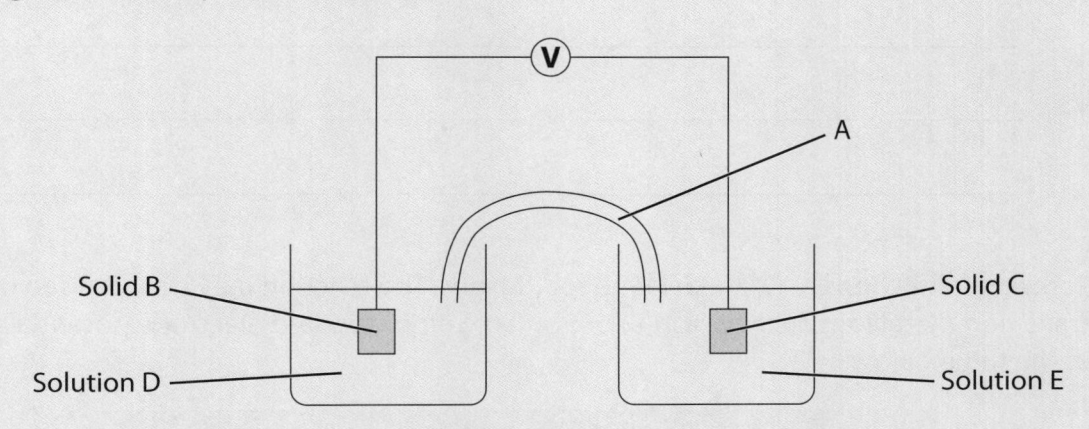

(a) What is represented by A and state two reasons why it is used in an electrochemical cell.

_____

_____

_____ [3]

(b) What are solids B and C?

Solid B _____

Solid C _____ [2]

(c) State the solute(s) in solutions D and E and the concentration of each solute.

Solution D _____

_____

Solution E _____

_____ [4]

(d) Write the conventional representation for the cell.

_____ [2]

(e) Calculate the EMF of the cell.

_____

_____ [2]

(f) Which electrode is the negative electrode in the cell? Explain your answer.

_____

_____

_____ [2]

7 Peroxodisulfate ions, $S_2O_8^{2-}$, react with iodide ions. The reaction may be catalysed by some transition metal ions such as iron(II) or iron(III). Some standard electrode potentials are given in the table below.

Half-equation	$E^\ominus$ /V
$S_2O_8^{2-}(aq) + 2e^- \rightleftharpoons 2SO_4^{2-}(aq)$	+2.01
$Fe^{3+} + e^- \rightleftharpoons Fe^{2+}$	+0.77
$Eu^{3+}(aq) + e^- \rightleftharpoons Eu^{2+}(aq)$	−0.35
$I_2(aq) + 2e^- \rightleftharpoons 2I^-(aq)$	+0.54

(a) Write an ionic equation for the reaction between peroxodisulfate ions and iodide ions.

_____ [1]

(b) Calculate the EMF for this reaction.

_____

_____ [2]

(c) Suggest why the rate of reaction is very slow.

_____

_____

_____ [2]

(d) The mechanism of catalysis with iron(II) ions is shown below.

Step 1: $S_2O_8^{2-} + 2Fe^{2+} \rightarrow 2SO_4^{2-} + 2Fe^{3+}$

Step 2: $2I^- + 2Fe^{3+} \rightarrow I_2 + 2Fe^{2+}$

(i) How does this mechanism show that iron(II) ions are a catalyst in the reaction.

_____

_____ [1]

(ii) Calculate the EMF of each step.

Step 1: _____

_____

Step 2: _____

_____ [2]

(iii) Suggest why iron(III) ions could also be a catalyst in this reaction.

_____

_____ [1]

(iv) Explain why Eu^{2+} ions cannot catalyst this reaction.

_____

_____

_____ [2]

8 Standard electrode potentials are measured relative to the standard hydrogen electrode. The standard hydrogen electrode potential is 0.00 V.

(a) Describe the standard hydrogen electrode.

_____

_____

_____

_____

_____ [4]

(b) Write an equation to represent the reduction process of the standard hydrogen electrode. Include state symbols.

_____ [2]

(c) Some standard electrode potentials are given below.

Half-equation	$E^\ominus$ / V
$Au^+ + e^- \rightleftharpoons Au$	+1.68
$O_2 + 4H^+ + 4e^- \rightarrow 2H_2O$	+1.23
$Fe^{3+} + e^- \rightleftharpoons Fe^{2+}$	+0.77
$Zn^{2+} + 2e^- \rightleftharpoons Zn$	−0.76
$Fe^{2+} + 2e^- \rightleftharpoons Fe$	−0.44
$SO_4^{2-} + 4H^+ + 2e^- \rightleftharpoons SO_2 + 2H_2O$	+0.17
$MnO_4^- + 8H^+ + 5e^- \rightleftharpoons Mn^{2+} + 4H_2O$	+1.51
$Cr_2O_7^{2-} + 14H^+ + 6e^- \rightleftharpoons 2Cr^{3+} + 7H_2O$	+1.33

(i) What would be observed if hydrogen gas was bubbled through a yellow solution containing iron(II) and iron(III) ions? Explain your answer using standard electrode potentials.

_____

_____

_____

_____ [3]

(ii) Write the conventional representation of the cell used to measure the standard electrode potential of the Zn^{2+}/Zn electrode.

_____ [2]

(iii) Explain, in terms of standard electrode potentials, why gold(I) ions reacts with water. Write an ionic equation for the reaction and state two observations you would make.

_____

_____

_____

_____

_____

_____ [5]

(d) Gold(I) ions are the strongest oxidising agent in the table in (c).

(i) What is meant by an oxidising agent?

_____

_____ [1]

(ii) Explain, in terms of standard electrode potentials, why gold(I) ions are the strongest oxidising agent in the table.

_____

_____ [2]

Commercial cells

9 The half equations and standard electrode potentials for the processes which are occurring in a lithium cell are shown below.

$Li^+ + e^- \rightarrow Li$ $E^\ominus = -3.03$ V

$Li^+ + CoO_2 + e^- \rightarrow Li^+[CoO_2]^-$ $E^\ominus = +0.57$ V

(a) Calculate the EMF of the lithium cell.

_____

_____ [2]

(b) Write an overall ionic equation for the process which occur during discharge.

_____ [1]

(c) Explain, using oxidation states, whether cobalt is oxidised or reduced during discharge.

_____
_____
_____ [3]

(d) Write an overall ionic equation for the reaction which occurs when the cell is charging.

_____ [1]

(e) Write the conventional cell representation of the lithium cell. Platinum is used in the positive electrode.

_____ [2]

(f) Suggest why water cannot be used as the solvent in a lithium cell.

_____
_____ [1]

10 The hydrogen fuel cell can operate in acidic or alkaline conditions. The conventional representation of the cell operating in alkaline conditions is:

$Pt|H_2(g)|OH^-(aq), H_2O(l)||O_2(g)|H_2O(l), OH^-(aq)|Pt$

The overall voltage of the cell is 1.23 V

(a) Write half equations for the oxidation and reduction processes occurring in the half cell.

Oxidation: _____

Reduction: _____ [2]

(b) The standard electrode potential for positive electrode is +0.40 V. Calculate the standard electrode potential for the negative electrode.

_____
_____ [2]

(c) Write an overall equation for the reaction which occurs in the hydrogen fuel cell. Include state symbols.

_____ [2]

(d) State two reasons why platinum is used for the electrodes.

_____

_____ [2]

(e) State two environmental benefits of the use of hydrogen fuel cells.

_____

_____

_____ [2]

11 The half equations and electrode potentials for the methanol fuel cell are:

Electrode 1: $CO_2(g) + 6H^+(aq) + 6e^- \rightarrow CH_3OH(l) + H_2O(l)$ $E^\ominus = +0.03$ V

Electrode 2: $6H^+(aq) + 1½O_2(g) + 6e^- \rightarrow 3H_2O(l)$ $E^\ominus = +1.23$ V

(a) Calculate the overall voltage which would be obtained from the methanol fuel cell.

_____

_____ [2]

(b) Write an overall equation for the reaction occurring in the methanol fuel cell.

_____ [2]

(c) Which electrode (1 or 2) would be the positive electrode? Explain your answer.

_____

_____ [1]

(d) State one disadvantage of fuel cells when compared to traditional cells.

_____

_____ [1]

12 Nickel-cadmium cells are used to power many different forms of electrical equipment. They are rechargeable cells which use nickel oxide hydroxide NiO(OH) and cadmium electrodes.

(a) Name one other type of rechargeable cell.

_____ [1]

(b) The overall equation for the reaction which occurs when a nickel cadmium cell is **charging** is:

$$2Ni(OH)_2 + Cd(OH)_2 \rightarrow 2NiO(OH) + 2H_2O + Cd$$

Explain how the oxidation states of cadmium and nickel changes during the charging reaction shown above.

_____

_____

_____ [2]

(c) The overall voltage delivered by the cell is 1.4 V.

One of the standard electrode potentials involved is:

$$Cd(OH)_2 + 2e^- \rightarrow Cd + 2OH^- \qquad E^\ominus = -0.88 \text{ V}$$

(i) Write a half equation for the oxidation and reduction processes which occur when the cell is **discharging**.

Oxidation: _____

Reduction: _____ [2]

(ii) Calculate the standard electrode potential of the nickel oxide hydroxide electrode.

_____

_____ [2]

(iii) State and explain which electrode is the negative electrode.

_____

_____

_____ [2]

(iv) Suggest why potassium hydroxide is used as the electrolyte is used in this cell.

_____

_____ [1]

5.7 Amines

Nomenclature and physical properties

1 Name each of the following compounds.

(a) CH₃CH₂NH₂	(b) 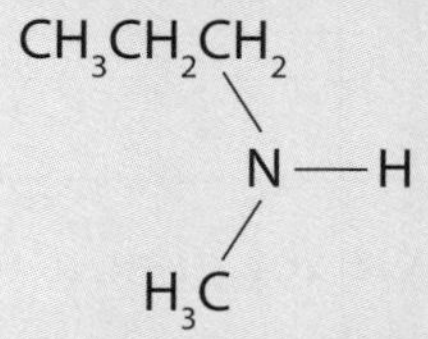 CH₃CH₂CH₂–NH–CH₃
(c) aniline (NH₂ on benzene ring)	(d) CH₃–NH–CH₂CH₃
(e) H₃C–CH(NH₂)–CH₃	(f) H₃C–CH(NH₂)–CH₂–CH(NH₂)–CH₃
(g) H₂NCH₂CH₂CH₂NH₂	(h) (CH₃CH₂)₂NH₂
(i) sec-butylamine (H₂N–CH(CH₃)–CH₂CH₃)	(j) CH₃CH₂CH₂–N(CH₂CH₃)(CH₂CH₃)

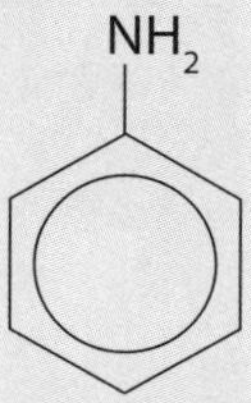

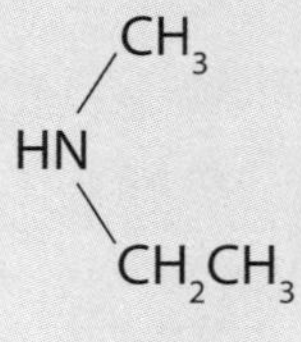

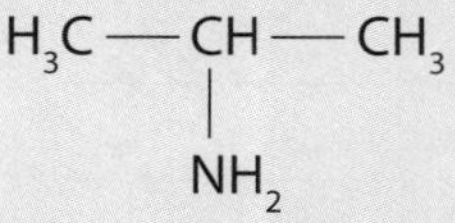

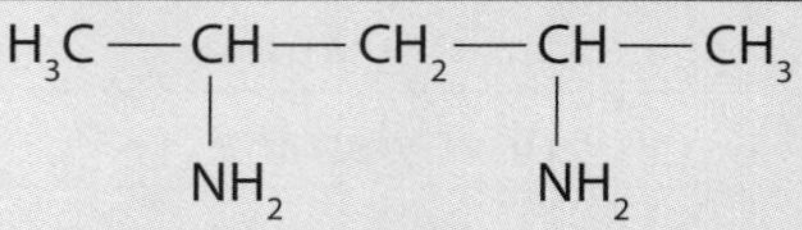

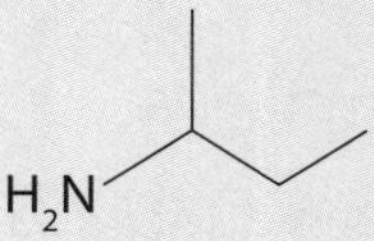

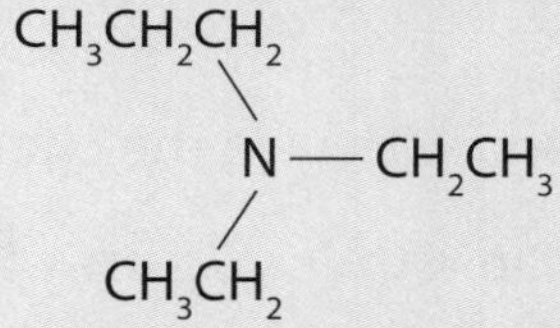

(a) _____   (b) _____

(c) _____   (d) _____

(e) _____   (f) _____

(g) _____   (h) _____

(i) _____   (j) _____

[10]

2 (a) Define the terms **primary amine** and **secondary amine**.

Primary amine: _____

_____

Secondary amine: _____

_____ [2]

(b) Classify the amines as primary, secondary or tertiary amines by placing a tick in each row in the table below.

Amine	1° (primary)	2° (secondary)	3° (tertiary)
methylamine			
ethylamine			
dimethylamine			
phenylamine			
triethylamine			

[5]

3 Ethylamine is used widely in the chemical industry and in organic synthesis.

(a) Draw and name the structure of an isomer of ethylamine which is a secondary amine.

_____ [2]

(b) Explain why the isomer drawn in (a) has a lower boiling point than ethylamine.

_____

_____ [2]

(c) Explain if ethylamine is soluble in water.

_____

_____ [2]

4 The primary amine A which has formula $CH_3CH_2CH_2NH_2$ is used as a solvent in organic synthesis and in the manufacture of drugs.

(a) Give the IUPAC name for A.

_____ [1]

(b) Draw the structure of a primary, secondary and tertiary amine which are isomers of A. Label each structure as primary, secondary or tertiary.

[3]

(c) Explain why the tertiary amine isomer has a lower boiling point than A.

_____

_____ [2]

Preparation of amines

5 (a) Ethanol and ammonia can react to form ethylamine and water, in the presence of a catalyst. Write an equation for this reaction.

_____ [1]

(b) Ethylamine can be prepared from bromoethane in one step.

(i) State the reagents and conditions used.

_____ [1]

(ii) Write an equation for this reaction.

_____ [1]

A2 2: ANALYTICAL, TRANSITION METALS, ELECTROCHEMISTRY AND ORGANIC NITROGEN CHEMISTRY

6 Write equations for the following reactions and name the organic product.

   (a) reduction of butanenitrile

   _____

   _____ [2]

   (b) 1-iodopropane and ammonia

   _____

   _____ [2]

   (c) reduction of nitrobenzene

   _____

   _____ [2]

7 (a) State the reagent and condition used in 6(a).

   _____

   _____ [2]

   (b) State the reagent(s) and condition used in 6(c).

   _____

   _____ [2]

8 Propylamine, $CH_3CH_2CH_2NH_2$ can be prepared by two different routes.

   Route 1: CH_3CH_2Br → compound C → $CH_3CH_2CH_2NH_2$

   Route 2: $CH_3CH_2CH_2Br$ → $CH_3CH_2CH_2NH_2$

   (a) Identify compound C.

   _____ [1]

   (b) Give the reagents for both stages in Route 1 and the single stage in Route 2.
   Route 1:

   _____

   _____

   Route 2:

   _____ [3]

5.7 AMINES

9 Below is a reaction scheme for the production of phenylamine.

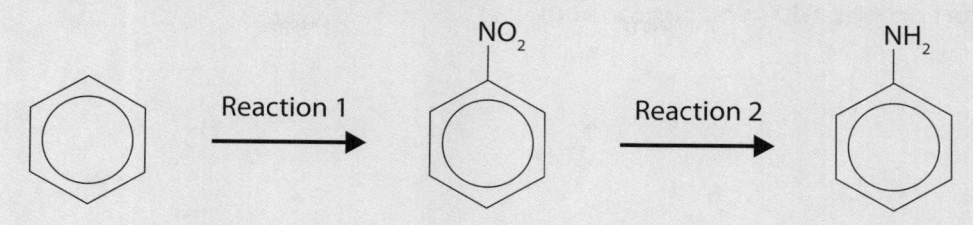

C₆H₆ →(Reaction 1)→ C₆H₅NO₂ →(Reaction 2)→ C₆H₅NH₂

(a) Give the reagents used to produce the electrophile needed in Reaction 1.

_____

_____ [1]

(b) Write an equation showing the formation of this electrophile and name the electrophile.

_____

_____ [2]

(c) (i) State the type of reaction which occurs in Reaction 2. Identify the reagents for Reaction 2.

_____

_____ [2]

(ii) Suggest why an aqueous solution is obtained in Reaction 2 even though phenylamine is insoluble in water and explain why sodium hydroxide is added at this stage.

_____

_____

_____ [2]

10 A compound W may be used to synthesise two different amines, Y and L.

$C_6H_5CH_3$ → compound X → $C_6H_4CH_3NH_2$
    W   Y

$C_6H_5CH_3$ → $C_6H_5CH_2Cl$ → $C_6H_5CH_2NH_2$
    W   K   L

(a) Give the IUPAC name for compound W.

_____ [1]

193

(b) W is heated with concentrated sulfuric and concentrated nitric acid. Draw the structure of the compound X which is formed.

[1]

(c) Name a suitable reducing agent to reduce X to phenylamine. Give an equation for the reduction using [H] to represent the reducing agent.

Reducing agent: _____

[2]

(d) Identify a reagent to convert K to L.

_____ [1]

Amines as bases

11 Write equations for the following reactions and name the organic product.

(a) ethylamine and hydrochloric acid

_____
_____ [2]

(b) methylamine and sulfuric acid

_____
_____ [2]

(c) phenylamine and nitric acid

_____
_____ [2]

(d) ethylammonium chloride and sodium hydroxide

_____
_____ [2]

(e) phenylamine and sulfuric acid

_____

_____ [2]

12 State and explain which one is the strongest base, in each of the following pairs.

(a) ethylamine or ammonia

_____

_____ [2]

(b) phenylamine or ammonia

_____

_____ [2]

(c) ethylamine or butylamine

_____

_____ [2]

13 State and explain the difference in base strength between phenylamine, propylamine and ammonia.

_____

_____

_____

_____

_____

_____ [4]

14 Which one of the compounds below is the weakest base?

A   ammonia

B   butylamine

C   methylamine

D   phenylamine [1]

15 The drug shown below is often used to form a salt which is prescribed as an antidepressant.

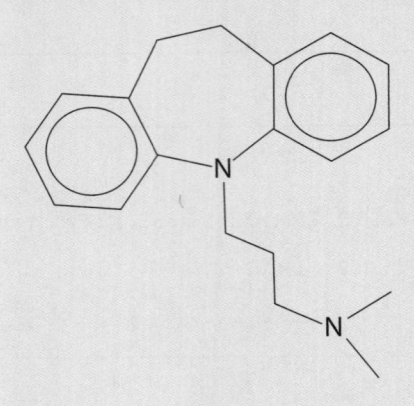

The salt is formed when one mole of the drug reacts with one mole of hydrochloric acid. Justify why the nitrogen atom in the ring is less likely to act as a base and be protonated than the other nitrogen atom, when the salt forms.

_____

_____

_____ [2]

16 The compound 1,3-dinitrobenzene can be converted to 1,3-diaminobenzene.

(a) State the reagents required to convert 1,3-dinitrobenzene to 1,3-diaminobenzene.

_____

_____ [2]

(b) Draw the structure of 1,3-diaminobenzene and of ethane-1,2-diamine.

1,3-diaminobenzene   ethane-1,2-diamine

[2]

(c) Explain why 1,3-diaminobenzene and ethane-1,2-diamine can act as bases.

_____

_____

_____

_____ [2]

(d) Explain why 1,3-diaminobenzene is a weaker base than ethane-1,2-diamine.

_____

_____

_____

_____ [4]

Reactions of amines

17 Write equations for the following reactions and name the organic product.

(a) phenylamine + ethanoyl chloride

_____ [2]

(b) ethylamine + propanoyl chloride

_____ [2]

(c) phenylamine + nitrous acid + hydrochloric acid to form benzene diazonium chloride

_____ [2]

(d) methylamine + nitrous acid

_____ [2]

(e) benzene diazonium chloride + phenol

_____ [2]

18 Phenylamine can form benzene diazonium chloride in a diazotisation reaction.

   (a) Write an equation for the diazotisation of phenylamine.

[2]

   (b) Explain, giving experimental details, how you would diazotise phenylamine.

_____
_____
_____ [3]

   (c) Explain what is meant by the term **coupling**.

_____
_____ [1]

   (d) Write an equation for the coupling of benzene diazonium chloride and phenol.

[2]

(e) State the condition for the reaction in 18(d) and give the appearance of the product

Condition _____

Appearance _____ [2]

19 The organic compound G is an azo dye that can be formed from 4-nitrophenylamine via the 4-nitrobenzene diazonium ion.

$O_2N-C_6H_4-N=N-C_6H_3(OH)-OH$

Compound G

(a) Draw the structure of the 4-nitrobenzene diazonium ion.

[1]

(b) Write the molecular formula for compound G and calculate its relative molecular mass.

_____

_____ [2]

(c) Calculate the mass of 4-nitrophenylamine (RMM (M_r) = 138) required to produce 28 g of dye G assuming a 75% yield with an excess of all other reagents. Give your answer to 2 significant figures.

_____

_____

_____

_____ [4]

(d) Circle the azo group on the structure. [1]

(e) Explain why dye G is coloured.

_____

_____

_____

_____ [3]

(f) State the names of the reagents and the reaction conditions for the formation of the 4-nitrobenzene diazonium ion from 4-nitrophenylamine.

_____

_____ [2]

20 1,4-diaminobutane $H_2N(CH_2)_4NH_2$ is produced by the breakdown of amino acids in the body.

(a) Explain why 1,4-diaminobutane is soluble in water.

_____

_____ [2]

(b) Write an equation for the reaction of 1,4-diaminobutane with excess ethanoyl chloride.

_____ [2]

(c) Explain how the purified product formed between 1,4-diaminobutane and excess ethanoyl chloride could be used to identify 1,4-diaminobutane.

_____

_____

_____ [2]

(d) Write an equation for the reaction of 1,4-diaminobutane with excess nitrous acid.

_____

_____ [2]

5.7 AMINES

21 Paracetamol may be produced from the reaction of 4-hydroxyphenylamine with ethanoyl chloride as shown by the equation below.

What mass of paracetamol is produced, assuming a 80% yield, if 10.9 g of 4-hydroxyphenylamine reacts with an excess of ethanoyl chloride?

A   12.1g

B   13.6g

C   15.1g

D   18.9g [1]

22 Benzene can react to form various organic chemicals, as shown in the reaction scheme below.

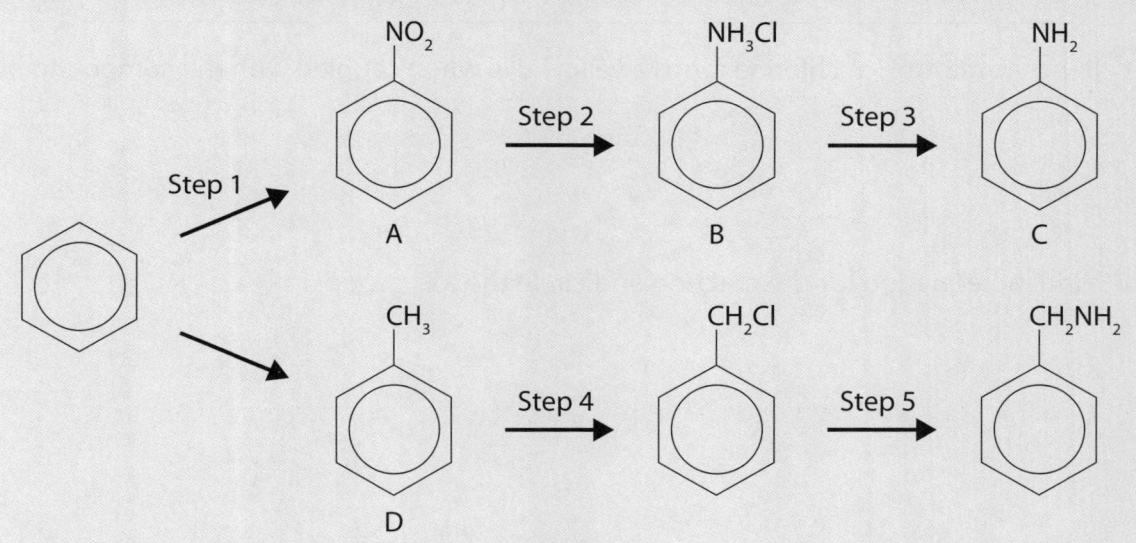

(a) Name compounds A, B, C and D.

A _____

B _____

C _____

D _____ [4]

(b) Complete the table for each of the steps 1, 2, 4 and 5 naming the type of reaction occurring and giving the reagents or the combination of reagents needed.

Step	Type of reaction	Reagents
1		
2		
4		
5		

[4]

(c) Name the reagent for Step 3.

_____ [1]

(d) Compound C is then converted to benzene diazonium chloride. Name the reagents and state the condition required for this conversion.

_____

_____

_____ [2]

(e) Benzene diazonium chloride forms a yellow dye when coupled with the compound below.

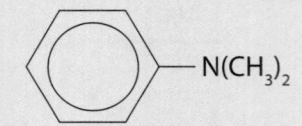

⟨benzene ring⟩—N(CH$_3$)$_2$

Write an equation for the reaction and circle the azo group.

[2]

5.8 Amides

1 Give the IUPAC names of the following compounds.

(a)

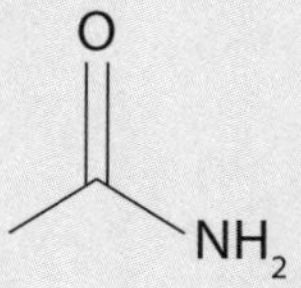

_____ [1]

(b)

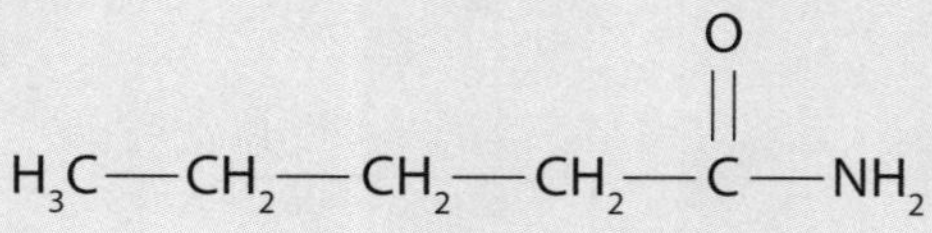

_____ [1]

(c)

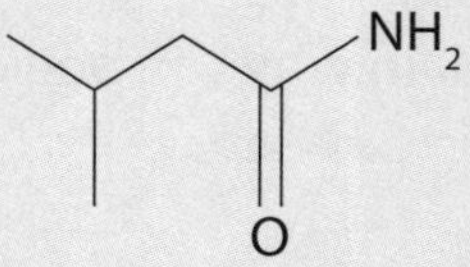

_____ [1]

(d)

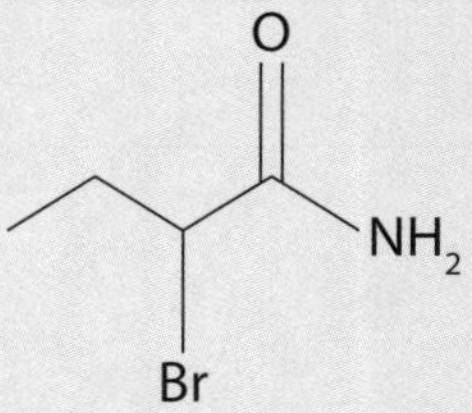

_____ [1]

(e)

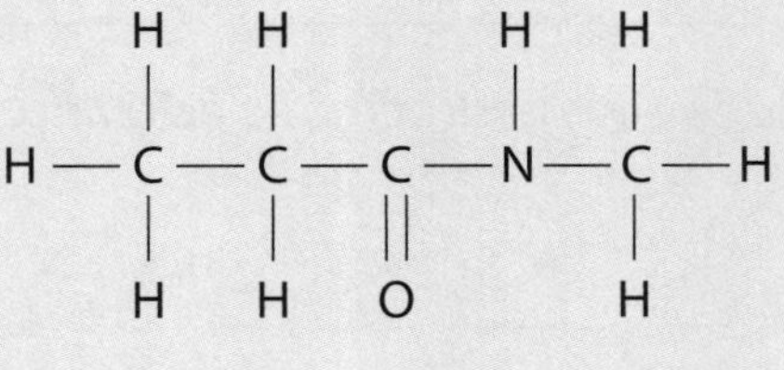

_____ [1]

(f)

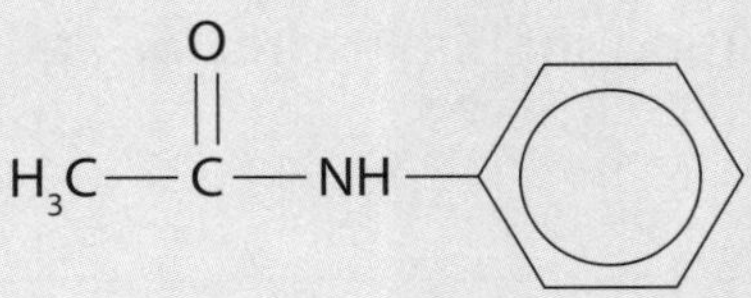

_____ [1]

(g) CH₃CH₂CONH₂

_____ [1]

(h)

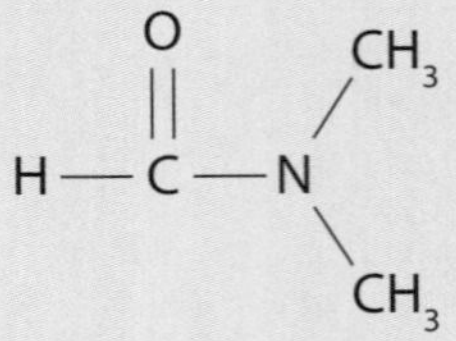

$$H-\underset{\|}{C}-N\begin{matrix}CH_3\\ \\CH_3\end{matrix}$$
$$O$$

_____ [1]

2 Explain why ethanamide is soluble in water.

_____

_____ [2]

3 Butane and propanamide have similar relative formula masses. Explain why the boiling point of propanamide is higher than that of butane.

_____

_____

_____ [2]

4 (a) Define the term **hydrolysis**.

_____

_____ [2]

(b) Write an equation for the hydrolysis of ethanamide using hydrochloric acid and name the organic product.

_____

_____ [2]

(c) Write an equation for the hydrolysis of ethanamide using sodium hydroxide and name the organic product.

_____

_____ [2]

5.8 AMIDES

(d) Write an equation for the hydrolysis of propanamide using sulfuric acid and name the organic product.

_____

_____ [2]

5 Amides can be dehydrated.

  (a) Explain the term **dehydration of amide**.

_____

_____ [1]

  (b) Name a dehydrating agent.

_____ [1]

  (c) Write an equation for the dehydration of propanamide and name the organic product.

_____

_____ [2]

6 Which one of the following is produced from the alkaline hydrolysis of ethanamide?

  A   Ammonia

  B   Ammonium chloride

  C   Ethanoic acid

  D   Water   [1]

7 Propanoic acid is used in preservatives in food.

  (a) Write an equation for the reaction of propanoic acid with ammonium carbonate and state the observations in this reaction.

  Equation: _____

  Observations: _____

_____ [4]

  (b) Write a word and symbol equation for the reaction which occurs when the salt formed in 7(a) is heated.

_____

_____ [2]

8 An amide X is produced by the reaction scheme below.

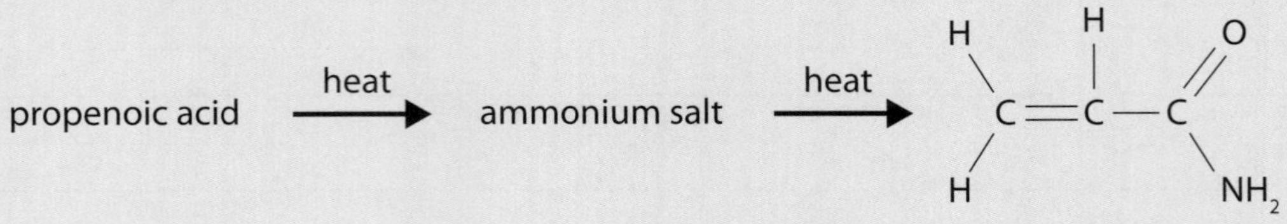

propenoic acid —heat→ ammonium salt —heat→ amide X (CH₂=CH–CONH₂)

(a) Draw the structure of propenoic acid showing all the bonds present.

[1]

(b) Suggest the formula and the name of the ammonium salt.

_____

_____ [2]

(c) Explain why the amide X shown above is soluble in water.

_____

_____ [2]

(d) The amide X produced by the above reaction scheme can be used to produce a nitrile.

  (i) Give the formula of the reagent used to convert X to a nitrile.

  _____ [1]

  (ii) Name the type of reaction taking place.

  _____ [1]

(e) Write the equations and state observations for the following reactions.

  (i) X + bromine water

  Equation: _____

  Observations: _____

  _____ [2]

(ii) X + sodium hydroxide

Equation: _____

Observations: _____

_____ [2]

9 Butanamide can be prepared from butanoic acid. Write two equations for this preparation.

_____

_____ [2]

10 The reaction scheme shows some reactions of propanamide.

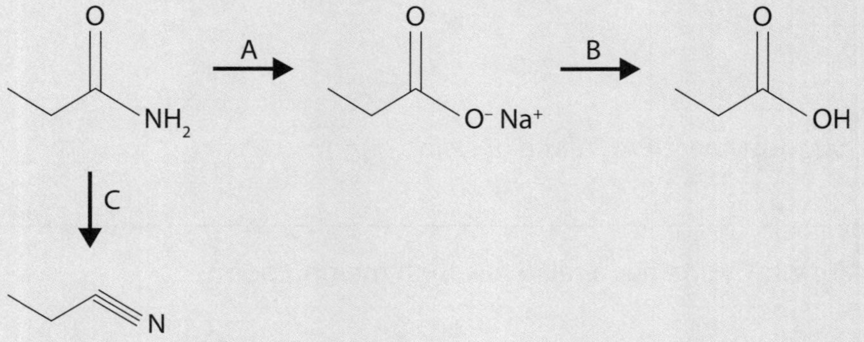

(a) Identify the reagents which could be used in steps A, B and C.

A _____

B _____

C _____ [3]

(b) Write an equation for the preparation of N-propylethanamide from the reaction of propylamine with ethanoyl chloride.

_____ [1]

5.9 Amino Acids

1 Lysine is an amino acid with the formula $H_2N(CH_2)_4CH(NH_2)COOH$. The compound is optically active.

   (a) (i) Explain the term optically active.

   _____

   _____ [2]

   (ii) Draw the 3D structure of lysine and label the chiral centre with a *.

   [2]

   (iii) Suggest the IUPAC name of lysine.

   _____ [1]

   (b) Explain why lysine has a relatively high melting point.

   _____

   _____ [2]

   (c) Write the formula of the organic ion present when lysine is dissolved in a solution of high pH.

   _____ [1]

   (d) Draw the structure of the product formed when lysine reacts with aqueous hydrochloric acid.

   [1]

5.9 AMINO ACIDS

(e) Draw the structure of a dimer formed when two molecules of lysine react.

[1]

(f) Draw the structure of two different dipeptides formed when one molecule of lysine reacts with one molecule of alanine.

[2]

(g) Another amino acid was found to have the following composition by mass.

Element	% composition
N	10.5
H	5.3
C	36.1
O	48.1

Deduce the empirical formula for this amino acid.

_____

_____

_____

_____ [2]

2 Valine and alanine are amino acids used in the synthesis of proteins. The structure of valine is shown below.

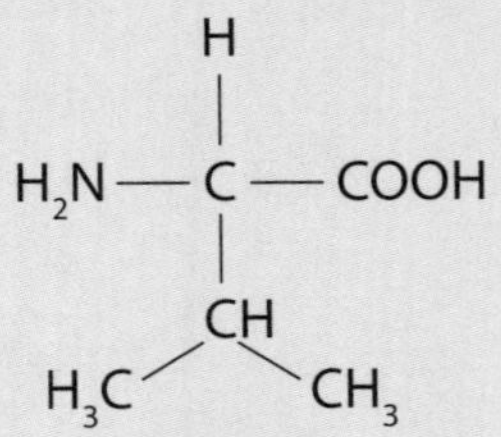

$$H_2N-\underset{\underset{\underset{H_3C}{\diagup}\overset{CH}{}\overset{\diagdown}{CH_3}}{|}}{\overset{H}{\underset{|}{C}}}-COOH$$

valine

(a) Draw the structure of alanine.

[1]

(b) Draw the structure of the zwitterion formed by valine.

[1]

(c) Valine is optically active. Draw the 3D representations of the optical isomers of valine.

[2]

(d) What is meant by the term **zwitterion**?

_____

_____ [2]

(e) Draw the structure of the zwitterions of the two dipeptides formed by alanine and valine.

[2]

(f) Draw the structure of the tripeptide formed when a valine molecule bonds to two alanine molecules, one on each side.

[2]

3 Which one of the following statements about alanine is **not** correct?

A   It has a relatively high melting point

B   It contains 32% carbon by mass

C   It exists as optical isomers

D   It is soluble in water

[1]

4 Consider the tripeptide shown below that is formed from three amino acids.

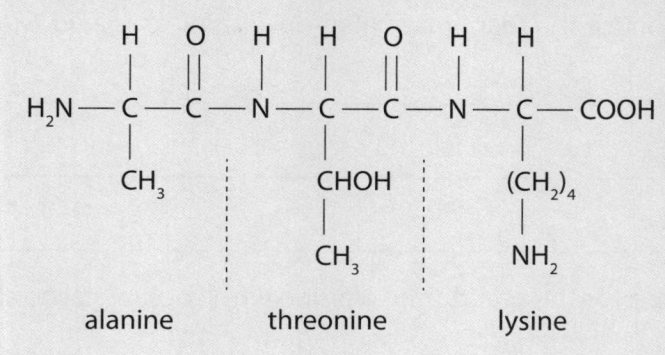

$$H_2N-\underset{\underset{CH_3}{|}}{\overset{\overset{H}{|}}{C}}-\overset{\overset{O}{\|}}{C}-\underset{}{\overset{\overset{H}{|}}{N}}-\underset{\underset{\underset{CH_3}{|}}{\overset{CHOH}{|}}}{\overset{\overset{H}{|}}{C}}-\overset{\overset{O}{\|}}{C}-\overset{\overset{H}{|}}{N}-\underset{\underset{\underset{NH_2}{|}}{\overset{(CH_2)_4}{|}}}{\overset{\overset{H}{|}}{C}}-COOH$$

alanine   threonine   lysine

(a) Name the process by which the tripeptide is split into three amino acids.

_____ [1]

(b) Give the IUPAC name for alanine.

_____ [1]

(c) Draw the structure of the zwitterion of threonine.

[2]

211

(d) Draw the structure of the species formed by alanine at low pH.

[1]

(e) Suggest the structure of the compound formed when alanine reacts with methanol in the presence of a small amount of concentrated sulfuric acid.

[1]

(f) Draw the structure and state the appearance of the compound formed when alanine reacts with copper(II) sulfate solution.

[1]

(g) Write an equation for the reaction of alanine with nitrous acid. State the observations.

Equation: _____

Observations: _____

_____ [2]

(h) Write an equation for the reaction of alanine with sodium carbonate.

Equation: _____

Observations: _____

_____ [2]

(i) Explain why alanine is soluble in water.

_____

_____ [1]

5 The diagram shown below shows a section of a primary protein structure.

(a) State the meaning of the term **primary protein structure**.

_____

_____ [1]

(b) Two other arrangements, A and B, which are part of the secondary structure of a protein are shown below.

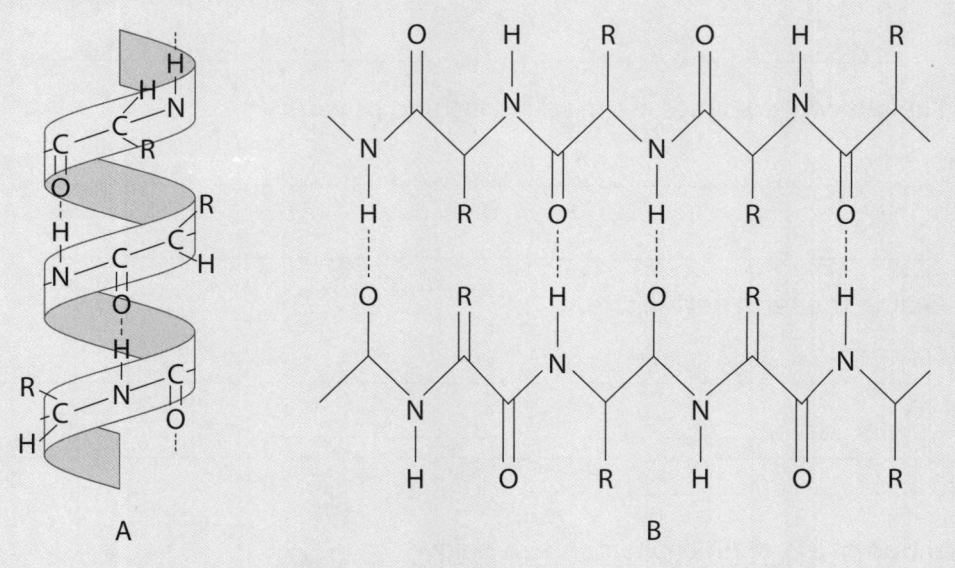

(i) Give the name of arrangement A and B.

A _____

B _____ [2]

(ii) Name the interaction represented by the dotted lines in both arrangements and explain how it arises.

_____

_____

_____ [2]

(c) Describe the tertiary structure of a protein.

_____

_____ [2]

(d) (i) Enzymes are proteins. Describe the action of an enzyme.

_____

_____ [2]

(ii) Explain the effect of temperature and pH on enzyme activity.

_____

_____

_____

_____ [4]

(iii) Explain why enzymes are used in washing powders.

_____

_____ [2]

(iv) Define the term **active site**.

_____

_____ [1]

6 A short section of a protein chain is shown below.

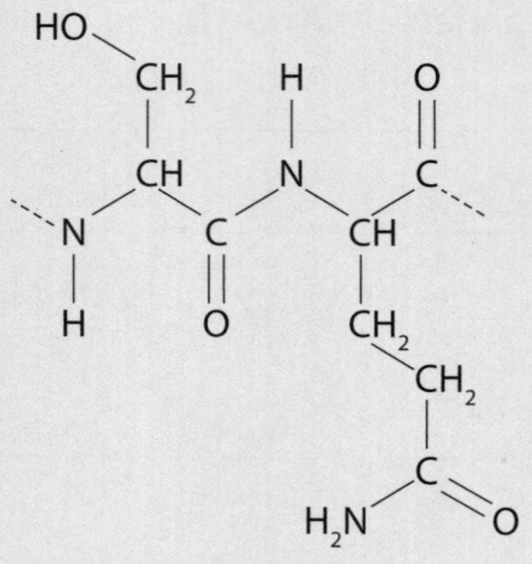

5.9 AMINO ACIDS

A student hydrolyses the protein with hot aqueous sodium hydroxide. Draw the structures of the organic products formed from this section of the protein.

[3]

7 A section of the primary structure of a protein is shown below.

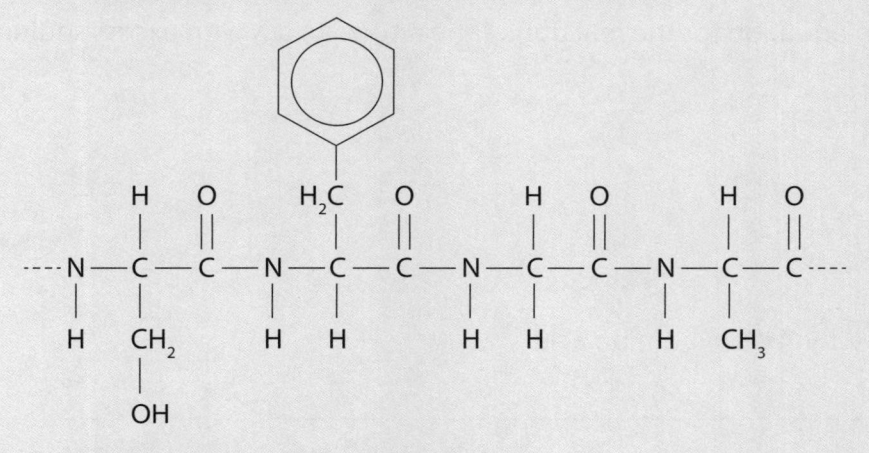

(a) State the number of monomers in this section.

_____ [1]

(b) Draw the structure of one of the monomer molecules used to form this section.

[1]

(c) Draw a circle around a peptide link in the diagram above. [1]

(d) Protein chains are often arranged in the shape of a helix. Name the type of interaction that is responsible for holding the protein chain in this shape.

_____ [1]

(e) Name the process by which this protein is split into amino acids and state a reagent used for the process.

_____

_____ [2]

8 An amino acid, X is shown below.

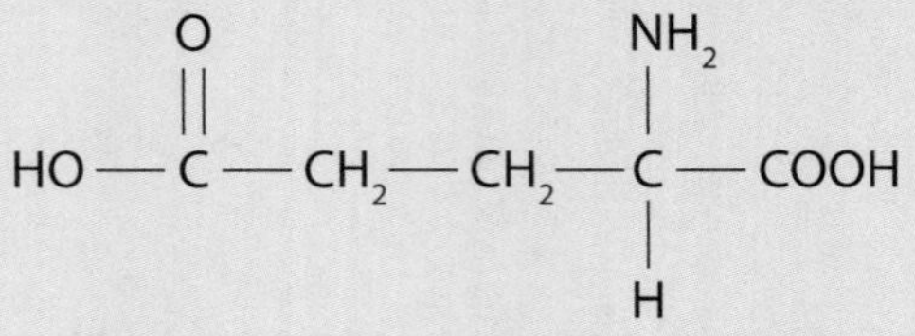

HO—C(=O)—CH₂—CH₂—C(NH₂)(H)—COOH

(a) Write an equation for the reaction of the amino acid X with excess sodium carbonate.

[2]

(b) Explain why X is an α amino acid.

_____

_____ [1]

(c) Give the IUPAC name of this amino acid.

_____ [1]

9 Some glycine is dissolved in a buffer solution of pH 11. What is the structure formed at this pH?

A   HOOCCH₂NH₃⁺

B   H₂NCH₂COOH

C   H₂NCH₂COO⁻

D   ⁻HNCH₂COO⁻   [1]

5.10 Polymers

1 Polymers are long chain molecules produced by addition or condensation reactions.

(a) (i) Nylon-6,6 is made from 1,6-diaminohexane and hexanedioic acid. Draw a section of the polymer showing two repeating units.

[2]

(ii) What type of condensation polymer in nylon-6,6?

_____ [1]

(iii) Explain what is meant by the term condensation polymer.

_____

_____ [2]

(iv) Draw the structure of a molecule which could be used instead of hexandioic acid to make nylon-6,6.

[1]

(b) The repeating unit of the polymer Terylene is shown below.

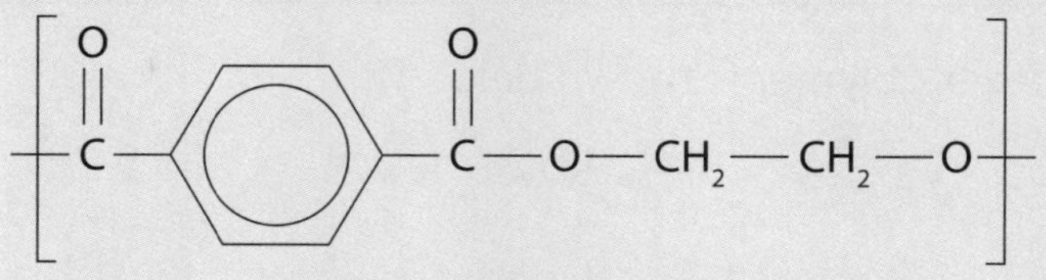

$$\left[\begin{array}{c} O \\ \parallel \\ C \end{array} - \bigcirc - \begin{array}{c} O \\ \parallel \\ C \end{array} - O - CH_2 - CH_2 - O \right]$$

(i) Draw the structure of the smaller of the two monomers used to make Terylene.

[1]

217

(ii) Name this monomer and state the type of polymerisation involved.

Name: _____

Type: _____ [2]

(c) Kevlar is a polymer used in bulletproof jackets. A section of the polymer chain is shown below.

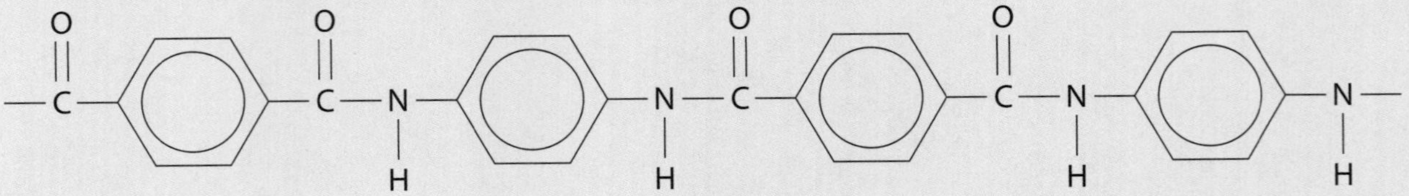

(i) State the number of repeating units shown.

_____ [1]

(ii) Draw the structures of the two monomers which can be used to produce Kevlar.

[2]

(iii) Explain if Terylene or Kevlar would have the higher melting point.

_____
_____
_____ [2]

(d) A section of another polymer is shown below.

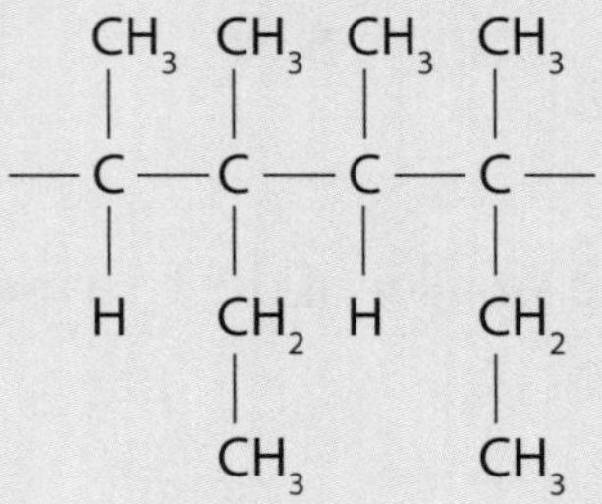

(i) Name the monomer used to produce this polymer.

_____ [1]

(ii) Name this type of polymer.

_____ [1]

(iii) State and explain an environmental advantage which Terylene has over the polymer in (d).

_____

_____ [2]

2 The compound $H_2N(CH_2)_4NH_2$ reacts with hexanedioic acid to form a condensation polymer.

(a) Suggest the IUPAC name for $H_2N(CH_2)_4NH_2$.

_____ [1]

(b) Draw the repeating unit of the condensation polymer formed when $H_2N(CH_2)_4NH_2$ reacts with hexanedioic acid.

[1]

(c) State and explain the nature of the intermolecular forces which occur between molecules of the condensation polymer.

_____

_____

_____ [2]

(d) Name the type of condensation polymer formed and suggest why it is biodegradeable.

_____

_____

_____ [2]

3 Terephthalic acid, HOOCC$_6$H$_4$COOH is obtained by the oxidation of 1,4-dimethylbenzene.

(a) Write an equation for the oxidation of 1,4-dimethylbenzene to terephthalic acid. Use [O] to represent the oxidising agent.

[1]

(b) Write an equation for the reaction of terephthalic acid to form its dimethyl ester.

[1]

(c) Terephthalic acid is used in the production of the polymer polyethylene terephthalate (PET).

(i) Draw the repeating unit of PET.

[1]

(ii) Explain why PET is biodegradable and suggest how the process could be speeded up.

_____

_____ [2]

4 (a) The repeating unit of a polymer is shown below.

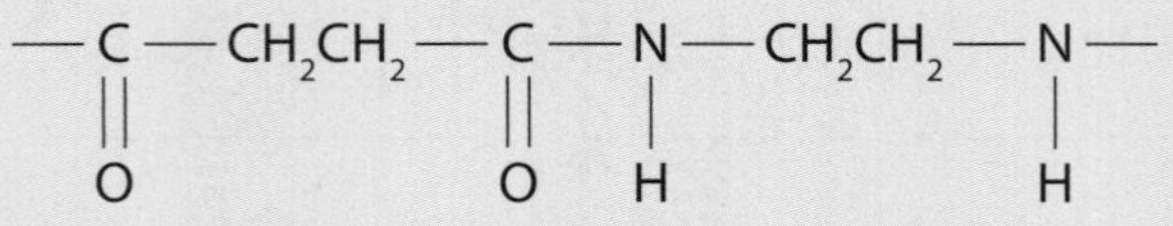

$$-\text{C}-\text{CH}_2\text{CH}_2-\underset{\text{O}}{\overset{\text{||}}{\text{C}}}-\underset{\text{H}}{\text{N}}-\text{CH}_2\text{CH}_2-\underset{\text{H}}{\text{N}}-$$

(with C=O on the left carbon)

Name the type of polymer and the type of polymerisation.

_____

_____ [2]

5.10 POLYMERS

(b) Part of a polymer chain is shown below.

$$-\underset{\underset{H}{|}}{\overset{\overset{H}{|}}{C}}-\underset{\underset{H}{|}}{\overset{\overset{CH_3}{|}}{C}}-\underset{\underset{CH_3}{|}}{\overset{\overset{H}{|}}{C}}-\underset{\underset{H}{|}}{\overset{\overset{H}{|}}{C}}-\underset{\underset{H}{|}}{\overset{\overset{H}{|}}{C}}-\underset{\underset{H}{|}}{\overset{\overset{CH_3}{|}}{C}}-\underset{\underset{CH_3}{|}}{\overset{\overset{H}{|}}{C}}-\underset{\underset{H}{|}}{\overset{\overset{H}{|}}{C}}-H$$

Name the monomer which produces this polymer.

_____ [1]

(c) HOCH$_2$CH$_2$CH$_2$CH$_2$COCl can react with itself to form a polyester. Draw the repeating unit of the polyester formed.

[1]

(d) Draw the repeating unit of the polymer formed from the monomer (CH$_3$)$_2$C=CHCH$_3$.

[1]

(e) In terms of the intermolecular forces between the polymer chains, suggest why polyamides can be made into strong fibres, whereas polyalkenes produce weaker fibres.

_____

_____ [2]

5 (a) Some sections of condensation polymers are shown below. For each part (i) to (iv) draw the structures of two monomers that could be used to form the polymer.

(i)

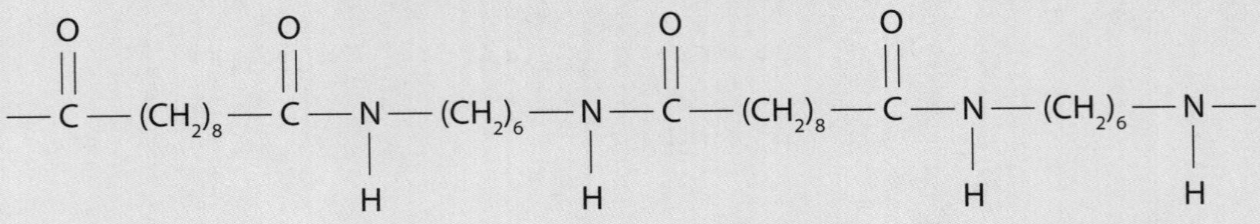

—C(=O)—(CH₂)₈—C(=O)—N(H)—(CH₂)₆—N(H)—C(=O)—(CH₂)₈—C(=O)—N(H)—(CH₂)₆—N(H)—

[2]

(ii)

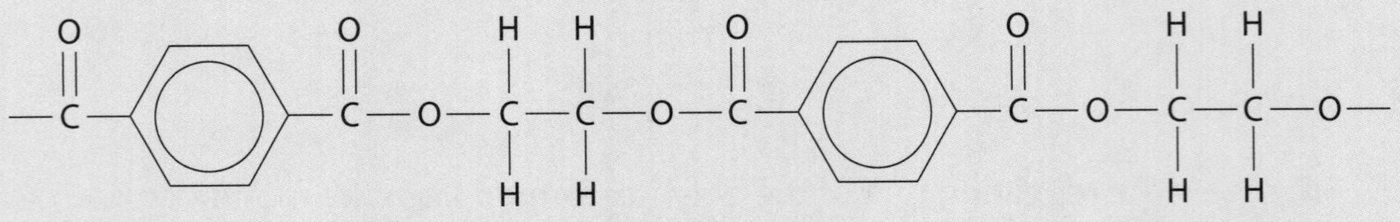

—C(=O)—C₆H₄—C(=O)—O—CH₂—CH₂—O—C(=O)—C₆H₄—C(=O)—O—CH₂—CH₂—O—

[2]

(iii)

CH₃—C(=O)—C₆H₄—C(=O)—N(H)—C₆H₄—N(H)—

[2]

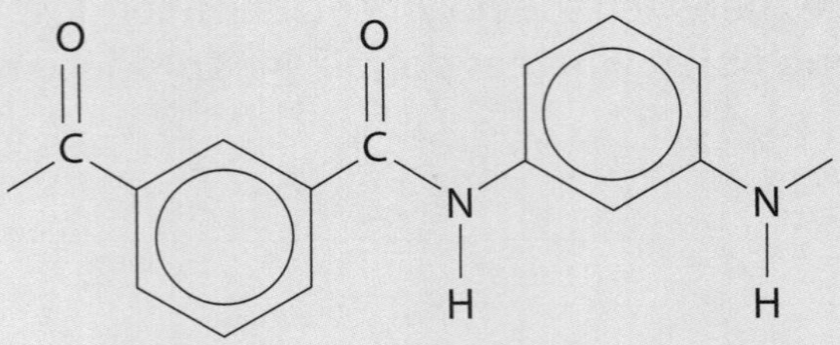

[2]

(iv)

[2]

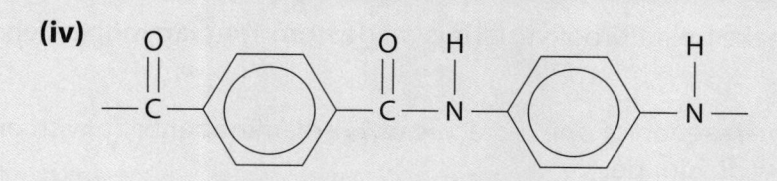

(b) Identify the polymers in (a) as polyesters or polyamides.

(i) _____ [1]

(ii) _____ [1]

(iii) _____ [1]

(iv) _____ [1]

6 The structure below shows a condensation polymer.

(a) Circle the repeating unit in the polymer structure. [1]

(b) Name the monomers from which this polymer is made.

_____

_____ [2]

(c) State the number of repeating units shown.

_____ [1]

7 Kevlar is a polymer used to make bullet proof vests. It is made from 1,4-diaminobenzene and terephthaloyl dichloride.

   (a) Write the equation for the reaction of one molecule of 1,4-diaminobenzene with one molecule of terephthaloyl dichloride.

   [2]

   (b) Explain why Kevlar is biodegradeable.

   _____

   _____ [2]

5.11 Chemistry in Medicine

1. Which one of the following ions is present in haemoglobin?
   A Cu^{2+}
   B Fe^{2+}
   C Fe^{3+}
   D Mg^{2+}
   [1]

2. Acid anhydrides are used in the synthesis of many pharmaceutical drugs including aspirin. Which one of the following is the correct skeletal formula of ethanoic anhydride?

   A

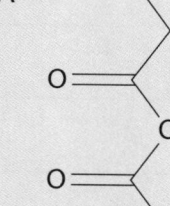

   B

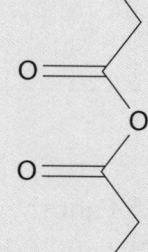

   C

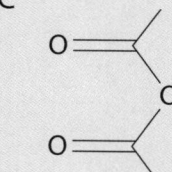

   D

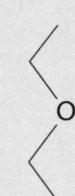

   [1]

3. Which one of the following may be used to treat eye infections in newborn children?
   A aspirin
   B cisplatin
   C salicylic acid
   D silver nitrate
   [1]

4 Magnesium hydroxide is often used to treat excess stomach acid.

(a) Suggest why barium hydroxide would not be used.

_____ [1]

(b) An antacid remedy contains magnesium hydroxide as its active component. 1.25 g of a sample of the antacid was reacted with 25.0 cm³ of 2.15 mol dm⁻³ hydrochloric acid (an excess). The resulting solution was diluted to 250.0 cm³ and 25.0 cm³ portions titrated against 0.0960 mol dm⁻³ sodium hydroxide solution. The mean titre was 19.4 cm³. Calculate the percentage by mass of magnesium hydroxide in the antacid remedy. Give your answer to 1 decimal place.

_____

_____

_____

_____

_____ [5]

(c) The manufacturers of another antacid remedy report that it contains 92.5 % calcium carbonate as its active ingredient. A 1.25 g sample was tested in the same way as the sample in (b). Calculate the mean titre which would have been obtained if the percentage given by the manufacturers is correct. Give your answer to 3 significant figures.

_____

_____

_____

_____ [5]

5 Skin has a pH of around 4.5 to 6.0. The acidity is caused by free fatty acids in the acid mantle such as sebacic acid which is shown below. An excess of acids can cause skin to become too acidic and oily.

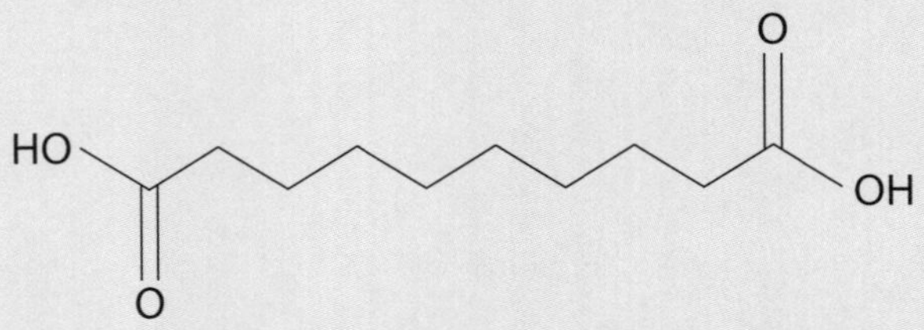

5.11 CHEMISTRY IN MEDICINE

(a) Suggest the IUPAC name for sebacic acid.

_____ [1]

(b) An excess of acids can cause the pH to drop below 4.5 which can cause skin irritation. Sebacic acid may be made more water soluble and removed by reaction with a weak alkali such as sodium carbonate.

  (i) Write an equation for the reaction of sebacic acid with excess sodium carbonate.

  _____ [1]

  (ii) Suggest why the product of the reaction with sodium carbonate is more water soluble than the acid itself.

  _____

  _____ [2]

(c) Warts may be removed using a 17% (17 g/100 g water) solution of salicylic acid.

  (i) Write the molecular formula for salicylic acid.

  _____ [1]

  (ii) A single dose of the wart treatment should contain 300 μmol of salicylic acid. Calculate the volume, in cm³, of 17% salicylic acid required to deliver this dose. The density of water is 1.00 g cm^{-3} and 1 μmol = 1 × 10^{-6} mol. Give your answer to 2 significant figures.

  _____

  _____

  _____ [3]

6 The structure of cis-diamminedichloropaltinium(II), often known as CDDP, is shown below.

$$H_3N \diagdown \diagup Cl$$
$$Pt$$
$$H_3N \diagup \diagdown Cl$$

(a) What is the common name used for this complex and what is it used for?

_____

_____ [2]

227

(b) Draw the structure of trans-diamminedichloroplatinum(II).

[1]

(c) Give the shape of the complex, its bond angle and its co-ordination number.

shape: _____

bond angle: _____

co-ordination number: _____ [3]

(d) The mode of action of CDDP targets the base guanine in DNA and prevents DNA replication.

   (i) What is meant by DNA replication?

   _____

   _____ [2]

   (ii) The first step in one of the modes of action of CDDP is where one chloro ligand is substituted for a water ligand. Write an equation for this ligand substitution.

   _____ [2]

   (iii) The base guanine is shown below.

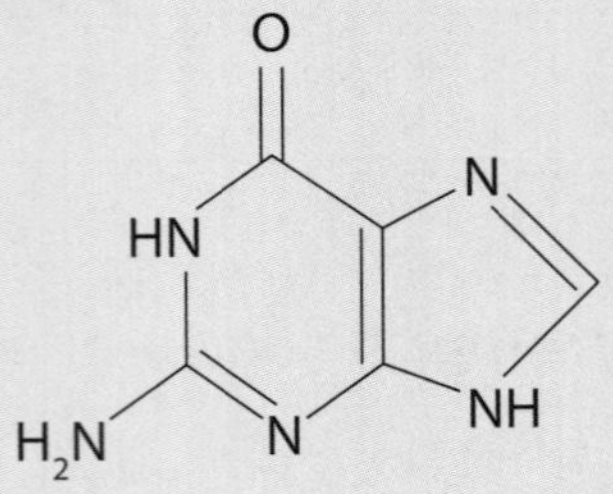

   Write the molecular formula for guanine.

   _____ [1]

(e) CDDP is often synthesised from potassium tetrachloroplatinate(II). Suggest a formula for potassium tetrachloroplatinate(II).

_____ [1]

5.11 CHEMISTRY IN MEDICINE

7 The structure of edta is shown below.

$$\text{HOOCH}_2\text{C}\diagdown\text{N}-\underset{\underset{\text{H}}{|}}{\overset{\overset{\text{H}}{|}}{\text{C}}}-\underset{\underset{\text{H}}{|}}{\overset{\overset{\text{H}}{|}}{\text{C}}}-\text{N}\diagup\text{CH}_2\text{COOH}$$
with HOOCH$_2$C and CH$_2$COOH also bonded to the two N atoms.

(a) Edta is manufactured from ethane-1,2-diamine using an aqueous solution of sodium cyanide and methanal. The tetrasodium salt is produced and ammonia. Write an equation for this reaction.

_____ [2]

(b) Blood collection tubes are often coated with a solution of the tripotassium salt of edta and the water is allowed to evaporate. Edta is a polydentate ligand. 1.5 mg of the tripotassium salt of edta are required for 1 cm³ of blood.

(i) Explain the role that potassium edta has in blood collection tubes.

_____

_____ [2]

(ii) Write the formula for tripotassium edta.

_____ [1]

(iii) Calculate the number of moles of tripotassium edta which would be required for a 4 cm³ sample of blood.

_____

_____

_____ [2]

(iv) What is meant by a **polydentate** ligand?

_____

_____ [2]

229

(c) Trisodium citrate may also be used in blood collection tubes. It may be formed from the reaction of citric acid (2-hydroxypropane-1,2,3-tricarboxylic acid) with excess sodium hydroxide solution.

(i) Draw the structure of citric acid.

[1]

(ii) Write an equation for the reaction of citric acid with excess sodium hydroxide.

_____ [2]

8   Haem is a complex of iron(II) with a porphyrin molecule. Its structure is shown below.

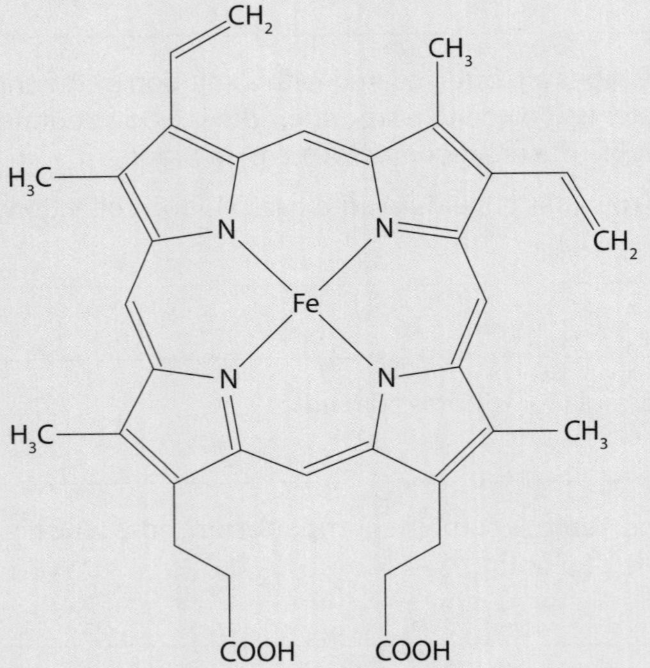

(a) What is the co-ordination number of the haem complex shown?

_____ [1]

(b) Explain how haem forms haemoglobin and transports oxygen in blood.

_____

_____

_____

_____ [3]

(c) In relation to haemoglobin, how is carbon monoxide poisonous?

_____

_____

_____

_____ [3]

9 Aspirin, shown below, reacts with sodium hydroxide. 1 mole of aspirin reacts with 2 moles of sodium hydroxide as the acid group reacts and the ester group undergoes hydrolysis. Aspirin content is often estimated using a back titration.

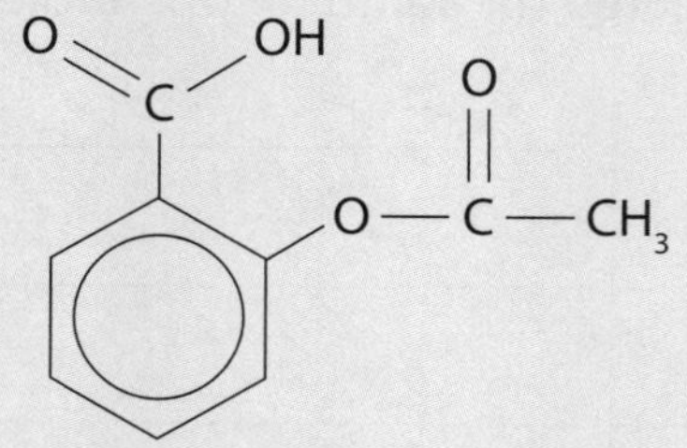

(a) Circle and label the ester group in aspirin. [1]

(b) Write an equation for the reaction of 1 mole of aspirin with 2 moles of sodium hydroxide.

_____ [2]

(c) Three aspirin tablets were crushed in a mortar using a pestle. They were added to 20 cm³ of ethanol and stirred. The solution formed was cloudy. The solution was mixed with 25.0 cm³ of 0.5 mol dm⁻³ sodium hydroxide (an excess) and warmed gently. The resulting solution was diluted to 250.0 cm³ in a volumetric flask using deionised water. A sample of 25.0 cm³ of this solution was titrated using 0.0125 mol dm⁻³ hydrochloric acid.

(i) Explain why the aspirin tablets were dissolved in ethanol rather than water.

_____

_____ [1]

(ii) Suggest why the solution formed with ethanol is cloudy.

_____

_____ [1]

(iii) Why was the solution warmed gently?

_____

_____ [1]

(iv) Calculate the expected mean titre if one tablet contains 300 mg of aspirin.

_____

_____

_____

_____

_____ [5]

(v) State an indicator which could be used for the titration and the colour change at the end point.

Indicator: _____

End point: _____ [2]

(vi) When the titration was carried out the titre was greater than the value expected. Suggest one reason why this could have happened.

_____

_____ [1]

5.11 CHEMISTRY IN MEDICINE

10 Alternatives to the drug cisplatin such as carboplatin are now used routinely in treating cancer. The structure of carboplatin is shown below. These drugs are platinum based complexes.

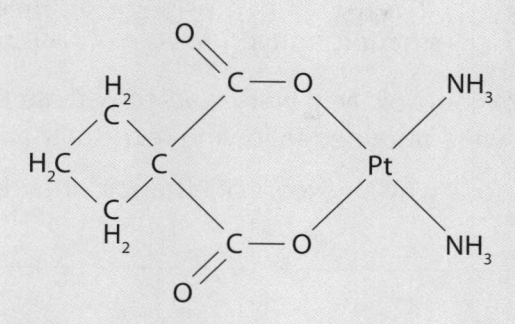

(a) Explain the mode of action of cisplatin in treating cancer.

_____

_____

_____

_____ [4]

(b) Explain why alternatives to cisplatin such as carboplatin have been developed.

_____

_____ [1]

(c) Write the formula of cisplatin.

_____ [1]

(d) Palladium based complexes such as the one shown below are also being developed to treat cancer. Suggest why palladium would be a suitable alternative to platinum.

_____

_____ [1]

11 The equation below shows the production of aspirin from 2-hydroxybenzoic acid and ethanoic anhydride.

$C_7H_6O_3$ + $C_4H_6O_3$ → $C_9H_8O_4$ + $C_2H_4O_2$
2-hydroxybenzoic acid   ethanoic anhydride   aspirin   ethanoic acid

(a) 0.500 g of 2-hydroxybenzoic acid were reacted with an excess of ethanoic anhydride. 485 mg of aspirin were obtained following recrystallisation from ethyl ethanoate.

(i) State two reasons why an excess of ethanoic anhydride was used.

_____
_____
_____ [2]

(ii) The density of ethanoic anhydride is 1.08 g cm^{-3}. Calculate the minimum volume of ethanoic anhydride required. Give your answer to 3 significant figures.

_____
_____
_____ [3]

(iii) Calculate the theoretical yield, in mg, of aspirin. Give your answer to 3 significant figures.

_____
_____
_____ [3]

(iv) Calculate the percentage yield of aspirin. Give your answer to 1 decimal place.

_____
_____ [1]

(v) Suggest one reason why the percentage yield is less than 100 %.

_____
_____ [1]

(b) Calculate the percentage atom economy for this synthesis of aspirin.

_____

_____ [1]

(c) Aspirin is often used as the sodium salt. Explain why the sodium soluble would make aspirin more effective.

_____

_____ [1]

12 Tablets of a calcium carbonate indigestion remedy were analysed as follows for their calcium carbonate content.

1. Crush five tablets and add to 50.0 cm³ of 1.50 mol dm⁻³ hydrochloric acid in a conical flask.

2. When effervescence has stopped, filter the mixture through filter paper into a 250 cm³ volumetric flask and the volume made up using deionised water.

3. Pipette 25.0 cm³ of the solution into a conical flask and titrate against 0.250 mol dm⁻³ sodium hydroxide solution using phenolphthalein indicator.

4. Repeat to obtain concordant results and calculate the mean titre.

(a) Suggest why the mixture was filtered.

_____

_____ [1]

(b) What is meant by concordant results?

_____

_____ [1]

(c) Write equations for the two reactions which occur.

_____

_____ [2]

(d) The mean titre was determined to be 13.1 cm³. Calculate the mass, in mg, of calcium carbonate in one tablet. Give your answer to an appropriate number of significant figures.

_____

_____

_____

_____

_____ [5]

13 GLC-MS is often used to separate components in a mixture and identify the components.

(a) What is meant by GLC-MS?

_____

_____ [1]

(b) Explain how GLC-MS may be used to separate and identify components in a mixture.

_____

_____

_____

_____

_____

_____ [5]

14 (a) The method below may be used to synthesise aspirin.

1. Add 5 cm³ of ethanoic anhydride (density = 1.08 g cm⁻³) to 2.0 g of salicylic acid in a conical flask.
2. Add 2 cm³ of concentrated phosphoric acid.
3. Heat for 20 minutes and stir using a magnetic stirrer.
4. Add 40 cm³ of deionised water to the flask and allow to cool.
5. Suction filter the crystals.
6. Rinse the crystals with cold water several times maintaining the suction.
7. Remove the crystals and add them to 25 cm³ of sodium hydrogencarbonate solution. Stir until effervescence stops.

5.11 CHEMISTRY IN MEDICINE

8. Suction filter again and wash the crystals with a little cold water twice whilst maintaining the suction.

9. Dissolve the crystals in a minimum volume of hot ethyl ethanoate and filter whilst hot.

10. Cool the filtrate and suction filter the crystals. Dry crystals in a desiccator.

(i) Write a structural equation for the formation of aspirin from ethanoic anhydride and ethanoic acid.

_____ [2]

(ii) Explain why ethanoic anhydride is used and not ethanoyl chloride or ethanoic acid.

_____
_____ [2]

(iii) What is the IUPAC name of salicylic acid?

_____ [1]

(iv) Write the formula of phosphoric acid and explain its role in this preparation.

_____
_____ [2]

(v) Explain why water was added in Step 4.

_____
_____ [1]

(vi) Why were the crystals washed with cold water in Step 6?

_____
_____ [1]

(vii) Why were the crystals placed in sodium hydrogencarbonate solution?

_____
_____ [1]

(viii) 2.24 g of aspirin were obtained. Calculate the percentage yield. Give your answer to 1 decimal place.

_____

_____

_____

_____

_____ [5]

(b) The preparation of aspirin may be monitored using thin-layer chromatography (TLC). The solvent used is ethyl ethanoate. Spots on the TLC plate may be viewed under UV light or an alkaline solution of potassium manganate(VII) may be used where any spots will appear yellow against the purple colour of the plate.

(i) Describe how you would prepare the TLC plate.

_____

_____

_____

_____ [3]

(ii) Explain how the TLC chromatogram is obtained and R_f values calculated.

_____

_____

_____

_____

_____

_____ [5]

(iii) Explain how you would know from the R_f values obtained from the chromatogram if the reaction was complete.

_____

_____ [2]

(iv) Suggest what type of reaction occurs between the organic compounds in the sample and the solution of potassium manganate(VII).

_____ [1]